CELESTIAL MECHANICS

By

HARRY POLLARD

THE

CARUS MATHEMATICAL MONOGRAPHS

Published by
THE MATHEMATICAL ASSOCIATION OF AMERICA

———

THE CARUS MATHEMATICAL MONOGRAPHS are an expression of the desire of Mrs. Mary Hegeler Carus, and of her son, Dr. Edward H. Carus, to contribute to the dissemination of mathematical knowledge by making accessible at nominal cost a series of expository presentations of the best thoughts and keenest researches in pure and applied mathematics. The publication of the first four of these monographs was made possible by a notable gift to the Mathematical Association of America by Mrs. Carus as sole trustee of the Edward C. Hegeler Trust Fund. The sales from these have resulted in the Carus Monograph Fund, and the Mathematical Association has used this as a revolving book fund to publish the succeeding monographs.

The expositions of mathematical subjects which the monographs contain are set forth in a manner comprehensible not only to teachers and students specializing in mathematics, but also to scientific workers in other fields, and especially to the wide circle of thoughtful people who, having a moderate acquaintance with elementary mathematics, wish to extend their knowledge without prolonged and critical study of the mathematical journals and treatises. The scope of this series includes also historical and biographical monographs.

© 1976 by

The Mathematical Association of America (Incorporated)
Library of Congress Catalog Card Number 76-51507

Complete Set ISBN 0-88385-000-1
Vol. 18 ISBN 0-88385-019-2

Printed in the United States of America

Current printing (last digit):

10 9 8 7 6 5 4 3 2 1

The Carus Mathematical Monographs

NUMBER EIGHTEEN

CELESTIAL MECHANICS

By

HARRY POLLARD

Purdue University

Published and Distributed by

THE MATHEMATICAL ASSOCIATION OF AMERICA

NOTE ON THE USE OF THIS BOOK

1. Vectors are printed in bold-face. Where possible the length of a vector is indicated by the same letter in italic. Thus the length of **v** is v. When this cannot be done, the length is indicated by the absolute-value symbol. Thus the length of $\mathbf{a} \times \mathbf{b}$ is $|\mathbf{a} \times \mathbf{b}|$.

2. Starred exercises are not necessarily difficult. The star indicates an important final result, or a result to be used later. Therefore starred exercises should not be omitted.

3. Unless otherwise stated, all references to formulas and exercises are made to the same chapter where they occur.

PREFACE

This is a corrected version of Chapters I–III of my *Mathematical Introduction to Celestial Mechanics* (Prentice-Hall, Inc., 1966). The acknowledgements made in the preface to that book apply equally well to this one. In addition, I am especially indebted to Professor D. G. Saari of Northwestern University for his thorough criticism of the original version.

HARRY POLLARD

To Helen and my family

CONTENTS

Chapter One

THE CENTRAL FORCE PROBLEM

1

Chapter Two

INTRODUCTION TO THE *n*-BODY PROBLEM   57

Chapter Three

INTRODUCTION TO HAMILTON-JACOBI THEORY   93

Index

CHAPTER 1

THE CENTRAL FORCE PROBLEM

1. FORMULATION OF THE PROBLEM

Celestial mechanics begins with the central force problem: to describe the motion of a particle of mass m which is attracted to a fixed center O by a force $mf(r)$ which is proportional to the mass and depends only on the distance r between the particle and O. The function f will be called a *law of attraction*. It is assumed to be continuous for $0 < r < \infty$.

Mathematically, the problem is easy to formulate. Indicate the position of the mass by the vector **r** directed from O. According to Newton's second law, the motion of the particle is governed by the equation

$$m\ddot{\mathbf{r}} = -mf(r)r^{-1}\mathbf{r},$$

where $r^{-1}\mathbf{r}$ is a unit vector directed to the position of the particle. If **v** denotes the velocity vector $\dot{\mathbf{r}}$, the equation can be written as the pair

$$\dot{\mathbf{r}} = \mathbf{v}, \qquad \dot{\mathbf{v}} = -f(r)r^{-1}\mathbf{r}. \qquad (1.1)$$

Observe that the value of m is irrelevant to the equations of motion. The problem is now this: to study the properties of pairs of vector-valued functions $\mathbf{r}(t)$, $\mathbf{v}(t)$ which simultaneously satisfy the Eqs. (1.1) over an interval of time.

The special case when the law of attraction is Newton's law of gravitation is the most important. In this case $f(r) = \mu r^{-2}$, where μ is a positive constant depending only on the units chosen and on the particular source of attraction. The Eqs. (1.1) become

$$\dot{\mathbf{r}} = \mathbf{v}, \qquad \dot{\mathbf{v}} = -\mu r^{-3}\mathbf{r}. \qquad (1.2)$$

2. THE CONSERVATION OF ANGULAR MOMENTUM: KEPLER'S SECOND LAW

Let us now assume that (1.1) is satisfied for some interval of time by the pair of functions $\mathbf{r}(t)$, $\mathbf{v}(t)$ which we write simply as $\mathbf{r}$, $\mathbf{v}$. From the second equation of the pair we conclude that

$$\mathbf{r} \times \dot{\mathbf{v}} = -f(r)r^{-1}(\mathbf{r} \times \mathbf{r}) = 0,$$

since the cross-product of a vector with itself is zero. Therefore, the derivative of the vector $\mathbf{r} \times \mathbf{v}$, which is $\mathbf{r} \times \dot{\mathbf{v}} + \mathbf{v} \times \mathbf{v}$, vanishes identically. Hence,

$$\mathbf{r} \times \mathbf{v} = \mathbf{c}, \qquad (2.1)$$

where $\mathbf{c}$ is a constant vector. The vector $m\mathbf{c}$ is called the *moment of momentum* and its length mc the *angular momentum* of the particle. We ignore these refinements and refer to either $\mathbf{c}$ or c as the angular momentum. The assertion (2.1) is known as *the conservation of angular momentum*.

An important consequence of the principle can be deduced immediately. According to (2.1) we have $\mathbf{c} \cdot \mathbf{r} = 0$. If $c \neq 0$, this means that $\mathbf{r}$ is always perpendicular to the fixed vector **c**. Consequently, if $c \neq 0$, *all the motion takes place in a fixed plane through the origin perpendicular to* **c**.

If $c = 0$, a little more subtlety is needed. Let **u** be a differentiable vector function of time and u its length. Since $u^2 = \mathbf{u} \cdot \mathbf{u}$, it follows that $u\dot{u} = \mathbf{u} \cdot \dot{\mathbf{u}}$. Therefore, if $u \neq 0$, we have

$$\frac{d}{dt} \frac{\mathbf{u}}{u} = \frac{u\dot{\mathbf{u}} - u\dot{u}}{u^2}$$

$$= \frac{(\mathbf{u} \cdot \mathbf{u})\dot{\mathbf{u}} - (\mathbf{u} \cdot \dot{\mathbf{u}})\mathbf{u}}{u^3},$$

or

$$\frac{d}{dt} \frac{\mathbf{u}}{u} = \frac{(\mathbf{u} \times \dot{\mathbf{u}}) \times \mathbf{u}}{u^3}, \qquad (2.2)$$

according to the vector formula

$$(\mathbf{a} \times \mathbf{b}) \times \mathbf{c} = (\mathbf{a} \cdot \mathbf{c})\mathbf{b} - (\mathbf{b} \cdot \mathbf{c})\mathbf{a}.$$

As an application of (2.2), let $\mathbf{u} = \mathbf{r}$. Then (2.2) becomes

$$\frac{d}{dt} \frac{\mathbf{r}}{r} = \frac{(\mathbf{r} \times \mathbf{v}) \times \mathbf{r}}{r^3} = \frac{\mathbf{c} \times \mathbf{r}}{r^3}, \qquad (2.3)$$

by (2.1). Therefore, if $c = 0$, the vector $\mathbf{r}/r$ is a constant, and *the motion takes place along a fixed straight line through the origin*.

In case $c \neq 0$, another important consequence can be deduced from (2.1). Introduce into the plane of motion a polar coordinate system centered at O and forming a right-handed system with the vector **c**. (See Fig. 1.) Then $\mathbf{r} = [r \cos \theta, r \sin \theta, 0]$ and $\mathbf{c} = [0, 0, c]$. A simple computation shows that (2.1) yields $r^2 \dot{\theta} = c$. According to the calculus, the rate at which area is swept out by a radius

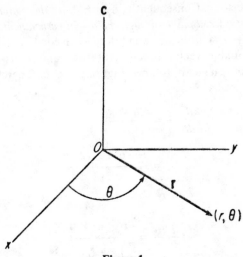

Figure 1

vector from O is just $\frac{1}{2} r^2 \dot{\theta}$. Therefore *the particle sweeps out area at the constant rate $c/2$*. This fact is Kepler's second law.

EXERCISE 2.1. Set up the equations of motion of a particle moving subject to two distinct centers of attraction, each with its own law of attraction.

EXERCISE 2.2. Suppose that a particle subject to attraction by a fixed center starts from rest, i.e., that at some instant $t = 0$ we have $v = 0$. Then by (2.1) $c = 0$ and the motion is linear. Suppose, moreover, that $f(r)$ is positive for $0 < r < \infty$. Prove that the particle must collide with the center of force in a finite length of time t_0.

EXERCISE 2.3. In the preceding problem, can you tell where the particle will be at each instant of time between 0 and t_0? First try the case $f(r) = \mu r^{-3}$ (inverse cube law), then $f(r) = \mu r^{-2}$ (inverse square law).

3. THE CONSERVATION OF ENERGY

So far we have found a vector **c** which remains constant throughout a particular motion. There is another constant of the motion which is of major importance, this time a scalar quantity called the *energy*. To find it, start with the second of Eqs. (1.1) and take the dot product of each side with **v**. We obtain

$$\dot{\mathbf{v}} \cdot \mathbf{v} = -f(r)r^{-1}(\mathbf{r} \cdot \mathbf{v})$$
$$= -f(r)r^{-1}r\dot{r}$$
$$= -f(r)\frac{dr}{dt}.$$

Integration of both sides yields

$$\tfrac{1}{2}v^2 = f_1(r) + h, \tag{3.1}$$

where $f_1(r)$ is a function whose derivative is $-f(r)$ and h is a constant. The function $f_1(r)$ is determined conventionally this way:

$$f_1(r) = \int_r^a f(x)dx$$

where (i) a is chosen as $+\infty$ if the integral converges; (ii) a is chosen to be 0 if the first choice leads to a divergent integral but the second does not; (iii) a is chosen to be 1 if the first two choices fail. Thus, if $f(r)$ is of the form $f(r) = \mu r^{-p}$, then $a = \infty$ if $p > 1$; $a = 0$ if $p < 1$; $a = 1$ if $p = 1$. The most important case is that of Newton:

$$f(r) = \mu r^{-2}, \qquad f_1(r) = \mu r^{-1}.$$

With the above convention the function $-mf_1(r)$ is known as the *potential energy* and is denoted by the symbol $-U$. The quantity $T = mv^2/2$ is called the *kinetic energy*, and $h_1 = mh$ the *energy*. The statement (3.1) becomes

$$T = U + h_1, \tag{3.2}$$

and is known as the *principle of conservation of energy*.

EXERCISE 3.1. Show that if $f(r) = \mu r^{-p}$, where $p > 1$, then a particle moving with negative energy cannot move indefinitely far from O.

EXERCISE 3.2. Show that if $f(r) = \mu r^{-p}$, then $f_1(r) = \mu(p-1)^{-1}r^{1-p}$ if $p \neq 1$ and $f_1(r) = \mu \log 1/r$ if $p = 1$.

*EXERCISE 3.3. Let $\mathbf{a} = \mathbf{r}$, $\mathbf{b} = \mathbf{v}$ in the standard vector formula

$$(\mathbf{a} \cdot \mathbf{b})^2 + (\mathbf{a} \times \mathbf{b})^2 = a^2 b^2.$$

Conclude that

$$v^2 = \dot{r}^2 + c^2 r^{-2}.$$

What is the physical meaning of the components $\dot{r}$ and $c/\mathbf{r}$ of $\mathbf{v}$? Show that the law of conservation of energy can be written

$$r^2 \dot{r}^2 + c^2 = 2r^2 [f_1(r) + h].$$

4. THE INVERSE SQUARE LAW: KEPLER'S FIRST LAW

In this section we shall assume that the particle is moving according to Newton's law of gravitation. The governing equations are then (1.2), which we repeat here for convenience as

$$\dot{\mathbf{r}} = \mathbf{v}, \qquad \dot{\mathbf{v}} = -\mu r^{-3}\mathbf{r}. \qquad (4.1)$$

It turns out that, in addition to the vector $\mathbf{c}$, there is another important vector which remains constant throughout the motion. It does not have a name in astronomical literature. We shall call it the *eccentric axis* and denote it by the symbol $\mathbf{e}$. To derive it, start with the formula (2.3) and multiply both sides by $-\mu$. Then

$$-\mu \frac{d}{dt} \frac{\mathbf{r}}{r} = \mathbf{c} \times (-\mu r^{-3}\mathbf{r}).$$

According to the second of Eqs. (4.1), this becomes

$$\mu \frac{d}{dt} \frac{\mathbf{r}}{r} = \dot{\mathbf{v}} \times \mathbf{c}.$$

Integration of both sides yields

$$\mu\left(\mathbf{e} + \frac{\mathbf{r}}{r}\right) = \mathbf{v} \times \mathbf{c}, \qquad (4.2)$$

where $\mathbf{e}$ is a constant of integration.

Since $\mathbf{r} \cdot \mathbf{c} = 0$, it follows that $\mathbf{e} \cdot \mathbf{c} = 0$. Hence, if $c \neq 0$, the vectors $\mathbf{e}$ and $\mathbf{c}$ are perpendicular, so that $\mathbf{e}$ lies in the plane of motion. If $c = 0$, $\mathbf{r}/r = -\mathbf{e}$, so that $\mathbf{e}$ lies along the line of motion; in this case the length e of $\mathbf{e}$ is always 1.

We shall now find the interpretation of e when $c \neq 0$. Take the dot product of both sides of (4.2) with $\mathbf{r}$. Then

$$\mu(\mathbf{e} \cdot \mathbf{r} + r) = \mathbf{r} \cdot \mathbf{v} \times \mathbf{c} = \mathbf{r} \times \mathbf{v} \cdot \mathbf{c} = \mathbf{c} \cdot \mathbf{c}.$$

Consequently,

$$\mathbf{e} \cdot \mathbf{r} + r = c^2/\mu. \qquad (4.3)$$

There are two cases. If $e = 0$, then $r = c^2/\mu$, a constant. Therefore the motion is circular. Moreover, according to the formula $r^2v^2 = r^2\dot{r}^2 + c^2$ of Ex. 3.3, it follows that $rv = c$, $v = \mu/c$, so that the particle moves with constant speed. By the law of conservation of energy, $v^2/2 = \mu/r + h$. Therefore $h = -\mu^2/2c^2$, a negative number. Observe finally that $2T = U$.

Suppose now that $e \neq 0$. In the plane of motion indicated by Fig. 1, introduce the vector **e** as shown in Fig. 2. The fixed angle from the x-axis to **e** will be denoted by ω. If (r, θ) represents a position Q of the particle, the angle $\theta - \omega$ will be denoted by f. The same position of the particle can then be represented as (r, f) if **e** is used as the axis of coordinates. It follows that $\mathbf{e} \cdot \mathbf{r} = er \cos f$ and Eq. (4.3) becomes

$$r = \frac{c^2/\mu}{1 + e \cos f} . \tag{4.4}$$

Consider the dotted line L in Fig. 2 drawn at a distance $c^2/\mu e$ from O, perpendicular to **e** and on the side of O to which **e** is directed. Equation (4.4), which can also be written $r = e\left(\dfrac{c^2}{\mu e} - r \cos f \right)$, simply says that the distance of the particle at Q from O is e times its distance from L. Consequently, *the particle moves on a conic section of eccentricity e with one focus at O*. This is Kepler's *first law*.

As (4.4) shows, the value of r is smallest when $f = 0$, since $e > 0$. Therefore the vector **e** is of length equal to the eccentricity and points to the position P at which the particle is closest to the focus.

There is some traditional terminology used by the astronomers that the reader ought to know. The position

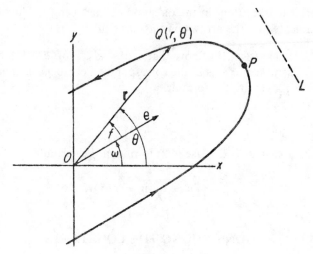

Figure 2

P is called the *pericenter*, the angle f the *true anomaly*. Various names are given to the pericenter, according to the source of attraction at O. If the source is the sun, P is called *perihelion*; if the earth, *perigee*; if a star, *periastron*. In the study of the solar system, the x-axis of Fig. 1 is fixed by astronomical convention. In that case, ω is the *argument of pericenter*.

We return to the geometry. The word *orbit* will be used to describe the set of positions occupied by the particle without any indication of the time at which a particular position is occupied. From the theory of conics it follows that if $0 < e < 1$ the orbit falls on an ellipse; if $e = 1$, on a parabola; and if $e > 1$, on a branch of hyperbola convex to the focus. Remember that in each case $c > 0$.

Since $r^2\dot\theta = c$ and $\dot\theta = \dot f$, it follows that $\dot f > 0$, so that the orbit is traced out in the direction of increasing f. This is indicated by the arrows on the curve in Fig. 2.

> *EXERCISE 4.1. Show that if $0 < e < 1$ or $e > 1$ the semi-major axis of the corresponding conic has length a given by the formula
>
> $$\mu a |e^2 - 1| = c^2.$$

> EXERCISE 4.2. Use (4.2) to obtain the formula
>
> $$\mu \mathbf{e} = \left(v^2 - \frac{\mu}{r} \right) \mathbf{r} - (\mathbf{r} \cdot \mathbf{v}) \mathbf{v}.$$

5. RELATIONS AMONG THE CONSTANTS

We pause at this point to remind the reader of some basic facts about differential equations. Let $f_i(z_1, \ldots, z_n)$, $i = 1, \ldots, n$ represent n functions with continuous first partial derivatives in some region of n-dimensional space, and let $(\zeta_1, \ldots, \zeta_n)$ be a particular point of this region. Then the system of differential equations

$$\dot{z}_i = f_i(z_1, \ldots, z_n), \qquad i = 1, \ldots, n \qquad (5.1)$$

will have a unique solution $z_i(t)$ defined in a neighborhood of $t = 0$, such that $z_i(0) = \zeta_i$, $i = 1, \ldots, n$.

Now consider the basic Eqs. (1.1) with the additional assumption that f has a continuous derivative. This includes the special cases $f(r) = \mu r^{-p}$. Each of the two Eqs. (1.1) stands in place of three scalar equations, so that the pair constitutes a system of order six of the form (5.1). Specifically, let x, y, z denote the components of $\mathbf{r}$ in a rectangular coordinate system and let α, β, γ denote the components of $\mathbf{v}$. The equations become

$$\dot{x} = \alpha$$

$$\dot{y} = \beta$$

$$\dot{z} = \gamma$$

$$\dot{\alpha} = -f(r)r^{-1}x$$

$$\dot{\beta} = -f(r)r^{-1}y$$

$$\dot{\gamma} = -f(r)r^{-1}z,$$

where $r^2 = x^2 + y^2 + z^2$. It follows that there is a unique solution satisfying six prescribed values of x, y, z, α, β, γ at $t = 0$. In vector form this says that the system (1.1) has a unique solution $\mathbf{r}(t)$, $\mathbf{v}(t)$ taking on prescribed values $\mathbf{r}_0$, $\mathbf{v}_0$ at time $t = 0$. These values can be prescribed arbitrarily.

In the special case $f(r) = \mu r^{-2}$, we have found that each of the quantities $\mathbf{c}$, $\mathbf{e}$, h remains constant during the motion and is therefore determined by its value at $t = 0$:

$$\mathbf{c} = \mathbf{r}_0 \times \mathbf{v}_0,$$

$$\mathbf{e} = \mu^{-1}(\mathbf{v}_0 \times \mathbf{c}) - r_0^{-1}\mathbf{r}_0,$$

$$h = \tfrac{1}{2}v_0^2 - \mu r_0^{-1}.$$

Since $\mathbf{c}$, $\mathbf{e}$, h constitute *seven* scalar quantities, it follows that there must be relations among them. We have already seen that there is a relation between $\mathbf{c}$ and $\mathbf{e}$, namely $\mathbf{c} \cdot \mathbf{e} = 0$. Therefore at most six of the seven quantities can be independent. Actually there is still another relation among the seven which reduces the number to *five*; it will be seen later that no further reduction is possible.

To obtain the new relation, square both sides of Eq. (4.2). Since v is perpendicular to c, we can replace $(\mathbf{v} \times \mathbf{c})^2$ by $v^2 c^2$ to obtain

$$\mu^2 \left(\mathbf{e} + \frac{\mathbf{r}}{r} \right)^2 = v^2 c^2$$

or

$$\mu^2 \left(e^2 + \frac{2}{r} \, \mathbf{e} \cdot \mathbf{r} + 1 \right) = v^2 c^2.$$

Replace v^2 by $2h + (2\mu/r)$ and $\mathbf{e} \cdot \mathbf{r}$ by $(c^2/\mu) - r$, according to Eq. (4.3). Then

$$\mu^2(e^2 - 1) = 2hc^2. \tag{5.2}$$

Notice that this agrees with the earlier results that $e = 1$ if $c = 0$ and $h = -\mu^2/2c$ if $e = 0$.

Equation (5.2) has the following important consequences. If $c \neq 0$, then $e < 1$, $e = 1$ or $e > 1$ according to whether the energy h is negative, zero, or positive. If $h \neq 0$ and $c \neq 0$ and a is the semi-major axis of the conic (see Ex. 4.1), then

$$a = \tfrac{1}{2} \, \mu |h|^{-1}. \tag{5.3}$$

From this and the energy formula $\tfrac{1}{2} v^2 = (\mu/r) + h$, we obtain these basic formulas:

$$v^2 = \mu \left(\frac{2}{r} + \frac{1}{a} \right) \qquad \text{if } h > 0;$$

$$v^2 = \frac{2\mu}{r} \qquad\qquad \text{if } h = 0; \tag{5.4}$$

$$v^2 = \mu \left(\frac{2}{r} - \frac{1}{a} \right) \qquad \text{if } h < 0.$$

These formulas still hold if $c = 0$ provided we adopt (5.3) as the definition of a; we shall do so.

EXERCISE 5.1.   What can you say about the orbit if $f(r) = -\mu r^{-2}$ rather than $f(r) = \mu r^{-2}$? This corresponds to a repulsive force rather than an attraction.

EXERCISE 5.2.   Use (5.4) to prove that in the case of elliptical motion the speed of the particle at each position Q is the speed it would acquire in falling to Q from the circumference of a circle with center at O and radius equal to the major axis of the ellipse.

*EXERCISE 5.3.   The area of an ellipse is $\pi a^2 (1 - e^2)^{1/2}$. We already know that if $c \neq 0$ the particle sweeps out area at the rate $c/2$. Combine these facts to show that if $0 < e < 1$ the period p of a particle, that is, the time it takes to sweep out the area once, is given by the formula $p = (2\pi/\sqrt{\mu})a^{3/2}$. This is Kepler's *third law*.

*EXERCISE 5.4.   Define the moment of inertia $2I$ by the formula $2I = mr^2$. Write $r^2 = (\mathbf{r} \cdot \mathbf{r})$ and prove that

$$\ddot{I} = 2T - U = T + h_1 = U + 2h_1.$$

In the case of circular motion I is constant so that $2T = U$, a result we already know from Sec. 4.

EXERCISE 5.5.   (*Hard.*) Use the preceding exercise to prove that if $c \neq 0$, $h > 0$, then $r/|t|$ approaches $\sqrt{2h}$ as $|t| \to \infty$. (The hypothesis $c \neq 0$ rules out the possibility of a collision with the origin in a finite time.)

6. ORBITS UNDER NON-NEWTONIAN ATTRACTION

The elegant method used in Sec. 4 to obtain orbits is essentially due to Laplace (who, however, did not have the vector concept available to him). It is applicable specifically to Newton's law of attraction. In the general case another method must be used. We know that if $c = 0$ the orbit is linear, so we shall assume that $c \neq 0$. Moreover, we assume that $f(r)$ has a continuous derivative.

Let us first dispose of the case of circular motion $r = r_0$. By the principle of conservation of energy, v is also a constant v_0 so the motion is uniform. The normal acceleration in the plane of motion is v_0^2/r_0 and this must be balanced by the attraction $f(r_0)$. Therefore, $v_0^2 = r_0 f(r_0)$. Since the velocity vector is perpendicular to the radius vector, it follows from $\mathbf{r} \times \mathbf{v} = \mathbf{c}$ that $rv = c$. Hence, $r_0 v_0 = c$, so that $c^2 = r_0^3 f(r_0)$. On the other hand, according to Ex. 3.3, the law of conservation of energy can be written

$$r^2 \dot{r}^2 + c^2 = 2r^2 \left[f_1(r) + h \right]. \tag{6.1}$$

Since $\dot{r} = 0$, we conclude that $c^2 = 2r_0^2[f_1(r_0) + h]$. Therefore, circular motion implies the two relations

$$c^2 = r_0^3 f(r_0), \qquad c^2 = 2r_0^2 \left[f_1(r_0) + h \right]. \tag{6.2}$$

Conversely, we shall show that if (6.2) holds for the value of r at some instant of time, say $t = 0$, then the particle moves uniformly in a circle of radius r_0. According to (6.1), the second of Eqs. (6.2) implies that $\dot{r}_0 = 0$.

We interrupt the argument at this point to obtain an important general formula. Starting with the equation $r^2 = \mathbf{r} \cdot \mathbf{r}$, we obtain $r\dot{r} = \mathbf{r} \cdot \mathbf{v}$ by differentiation. Another differentiation yields $r\ddot{r} + \dot{r}^2 = (\mathbf{r} \cdot \dot{\mathbf{v}}) + (\mathbf{v} \cdot \mathbf{v}) = (\mathbf{r} \cdot \dot{\mathbf{v}}) + v^2$. But (see Ex. 3.3) $v^2 = \dot{r}^2 + c^2 r^{-2}$, so that $r\ddot{r} = (\mathbf{r} \cdot \dot{\mathbf{v}}) + c^2 r^{-2}$. Since $\dot{\mathbf{v}} = -f(r)r^{-1}\mathbf{r}$, we have $(\mathbf{r} \cdot \dot{\mathbf{v}}) = -f(r)r^{-1}\mathbf{r} \cdot \mathbf{r} = -rf(r)$. Therefore $r\ddot{r} = -rf(r) + c^2 r^{-2}$, or, finally,

$$\ddot{r} - c^2 r^{-3} = -f(r). \tag{6.3}$$

We resume the argument. According to the first of Eqs. (6.2), Eq. (6.3) has the constant solution $r = r_0$. Moreover, since the values of r and $\dot{r}$ at $t = 0$ are given, the uniqueness theorem described in Sec. 5 tells us that this is the only possible solution. This completes the case of circular motion.

In the general case it is customary to start with (6.3) and remove the dependence on time by substitution from $r^2\dot{\theta} = c$. Specifically, let $r = \rho^{-1}$. Then $\dot{r} = -\rho^{-2}\dot{\rho} = -\rho^{-2}\rho'\dot{\theta} = -\rho^{-2}\rho'cr^{-2} = -c\rho'$, where the prime ($'$) denotes differentiation with respect to θ. Hence $\ddot{r} = -c\rho''\dot{\theta} = -c^2\rho''\rho^2$. Equation (6.3) becomes

$$\rho'' + \rho = c^{-2}\rho^{-2}f\left(\frac{1}{\rho}\right). \tag{6.4}$$

In general, this cannot be solved for ρ in terms of θ in any recognizable form and we content ourselves with some special cases.

Suppose first that $f(r) = \mu r^{-2}$, the Newtonian case. Then $\rho'' + \rho = \mu/c^2$. It follows that ρ has the form $(\mu/c^2) + A\cos\theta + B\sin\theta$ and its reciprocal r has the form demanded by (4.4), since $f = \theta - \omega$.

Another easy case is $f(r) = \mu r^{-3}$. Then $\rho'' + \rho = \mu c^{-2}\rho$ or $\rho'' + (1 - \mu c^{-2})\rho = 0$. The solutions of this are well known.

EXERCISE 6.1.   Classify the solutions in the case $f(r) = \mu r^{-3}$ according to the sign of $1 - \mu c^{-2}$. What if $1 - \mu c^{-2} = 0$?

EXERCISE 6.2. Show that for the direct first power law, $f(r) = \mu r$, the orbits are ellipses with center (*not* focus) at the origin.

EXERCISE 6.3. If we write Eq. (6.3) in the form $\ddot{r} - r\dot{\theta}^2 = -f(r)$, what is the physical meaning?

7. POSITION ON THE ORBIT: THE CASE $h = 0$

We return to the problem of motion under Newtonian attraction. It was shown in Sec. 5 that a knowledge of initial values $\mathbf{r}_0$, $\mathbf{v}_0$ determine the motion completely. In particular, these values give us $\mathbf{c}$ and $\mathbf{e}$, which by Secs. 2 and 4 determine the orbit. But there is still something missing: where is the particle located on its orbit at a prescribed time t_1?

It would be desirable to answer this question by giving the position $\mathbf{r}(t)$ as some explicit recognizable function of time. This is difficult to do directly. Instead, we adopt another procedure. We shall change from the original time t to a fictitious "time" u by a change of variable $t = t(u)$. If this change of variable is suitably chosen, it is easy to locate the particle for a prescribed value u_1 of u. In order to locate the particle at the real time t_1, it will be necessary to solve the equation $t_1 = t(u_1)$ for the corresponding value of u_1. With the choice of $t(u)$ made in this chapter, the variable u is called by the astronomers the *eccentric anomaly*.

We start with Eq. (6.1), remembering that in the case of Newtonian attraction the function $f_1(r)$ is μ/r. Then

$$(r\dot{r})^2 + c^2 = 2(\mu r + hr^2). \tag{7.1}$$

It will be supposed that u is chosen in such a fashion that $r\dot{u}$ is a constant k. Specifically, let

$$u = k \int_T^t \frac{d\tau}{r(\tau)} . \tag{7.2}$$

where k and T will be selected later. It is remarkable that the change of variable involves the still unknown function $r(t)$, but this will take care of itself. Since

$$\dot{r} = \frac{dr}{du} \dot{u} = \frac{dr}{du} kr^{-1},$$

Eq. (7.1) becomes

$$k^2(r')^2 + c^2 = 2(\mu r + hr^2), \tag{7.3}$$

where now the prime (') denotes differentiation with respect to u.

The treatment of this equation depends on the sign of h. In this section we confine ourselves to the case $h = 0$. With the choice $k^2 = u$, Eq. (7.3) then reads

$$(r')^2 + \frac{c^2}{\mu} = 2r. \tag{7.4}$$

If we differentiate both sides, the result is $r'r'' = r'$. Therefore, since r' cannot vanish over an interval (or r would be a constant!), it follows that $r'' = 1$. Therefore r is a quadratic in u which we write $r = \frac{1}{2}(u - u_0)^2 + A$. Substitution into (7.4) shows that $A = c^2/2\mu$. Moreover, since u is unspecified within an arbitrary constant by (7.2), we may choose $u_0 = 0$. Then

$$r = \frac{1}{2}\left(u^2 + \frac{c^2}{\mu}\right).$$

According to (7.2), $du/dt = k/r$, or $r\,du = k\,dt$. Moreover $u = 0$ when $t = T$. Therefore

$$k \int_T^t dt = \int_0^u r\,du$$

$$= \tfrac{1}{2} \int_0^u \left(u^2 + \frac{c^2}{\mu} \right) du,$$

or, because $k^2 = \mu$,

$$\sqrt{\mu}\,(t - T) = \tfrac{1}{6} u^3 + \frac{c^2}{2\mu}\,u.$$

In summary,

$$\sqrt{\mu}\,(t - T) = \tfrac{1}{6} u^3 + \frac{c^2}{2\mu}\,u,$$

$$r = \tfrac{1}{2} \left(u^2 + \frac{c^2}{\mu} \right). \tag{7.5}$$

Observe that, by the first equation of the pair, t is a strictly increasing function of u. This means that this equation can be solved uniquely for u in terms of t. Call this solution $u(t)$. Then $r = \tfrac{1}{2}[(u(t))^2 + (c^2/\mu)]$. It is easily verified that this satisfies the differential equation (7.1) when $h = 0$.

For the interpretation of T, it is best to separate the case $c \neq 0$, and $c = 0$. If $c \neq 0$ and $h = 0$, then $e = 1$, and we obtain for the orbit the parabola

$$r = \frac{c^2/\mu}{1 + \cos f}. \tag{7.6}$$

The smallest value of r is $c^2/2\mu$ and is achieved when $f = 0$. But this is the value of r when $u = 0$, or $t = T$. Therefore T is the time at which the particle is closest to the origin; it is called the time of *pericenter passage*. It can occur either before or after the initial time $t = 0$, but, since $\dot{f} > 0$, it can occur only once.

If $c = 0$, the equations read

$$6\sqrt{\mu}\,(t - T) = u^3; \qquad r = \tfrac{1}{2}u^2. \tag{7.7}$$

Therefore the time $t = T$ corresponds to collision with the origin. It must occur at some time. If $T > 0$, then it occurs after the initial time; the motion after the time T is no longer governed by the original equations, and we can talk about the motion only in the time interval $-\infty < t < T$. If $T < 0$, then the particle has been "emitted" from O at the time $t = T$ and we can speak of the motion only in the interval $T < t < \infty$.

To locate the position of the particle at time t, given $\mathbf{r}_0$ and $\mathbf{v}_0$, we proceed as follows. By the second of Eqs. (7.5), $\dot{r} = u\dot{u} = ukr^{-1}$. Therefore $r\dot{r} = (\mathbf{r} \cdot \mathbf{v}) = \sqrt{\mu}\,u$. Then the value u_0 at $t = 0$ is given by $\sqrt{\mu}\,u_0 = (\mathbf{r}_0 \cdot \mathbf{v}_0)$. Now let $t = 0$, $u = u_0$ in the first of Eqs. (7.5). This determines T. In order to find r for a given value of t we work backwards. Solve the first of Eqs. (7.5) for $u = u(t)$ and substitute into the second.

There are now two cases. If $c = 0$, then this knowledge of r determines the position completely since the line $\mathbf{e}$ containing the motion is known. On the other hand, if $c \neq 0$, it follows from (7.6) that there are two possible values of f for each value of r. It is clear that we must take f positive if $t > T$, f negative $t < T$; alternatively, $f > 0$ if $u > 0$, $f < 0$ if $u < 0$. The coordinates (r, f) then locate the particle completely.

EXERCISE 7.1. There is a standard formula from algebra for solving the cubic in (7.5) for $u = u(t)$. Write out the solution explicitly.

EXERCISE 7.2. Excluding the cases of collision, show that if $h = 0$, then $r|t|^{-2/3} \to (\frac{9}{2}\mu)^{1/3}$ as $|t| \to \infty$. Compare this with the corresponding result in case $h > 0$. (See Ex. 5.5.)

EXERCISE 7.3. Show that $u = (c/\sqrt{\mu})\tan f/2$, thus relating the two anomalies in the case $c \neq 0$. Hint: equate r as given by (7.5) and by (7.6).

8. POSITION ON THE ORBIT: THE CASE $h \neq 0$

If $h \neq 0$, there are these possible motions: linear if $c = 0$, hyperbolic if $c \neq 0$, $h > 0$, and elliptic if $c \neq 0$, $h < 0$. We now turn to the problem of location on the orbit at a prescribed time t.

Once again we start with the Eqs. (7.3) with the independent variable u defined by (7.2). This time we choose $k^2 = 2|h|$, or according to (5.3), $k^2 = u/a$. On division by k^2, Eq. (7.3) becomes

$$(r')^2 + \frac{ac^2}{\mu} = 2ar + \sigma(h)r^2,$$

where $\sigma(h) = 1$ if $h > 0$, $\sigma(h) = -1$ if $h < 0$. Add $\sigma(h)a^2$ to both sides and use the fact that $c^2/\mu = a(e^2 - 1)\sigma(h)$, as in (5.2). We obtain

$$(r')^2 + a^2e^2\sigma(h) = \sigma(h)\left[a + \sigma(h)r\right]^2.$$

Now define a new function $\rho(u)$ by

$$ea\rho = a + \sigma(h)r. \tag{8.1}$$

This converts the preceding equation for r' into

$$(\rho')^2 - \sigma(h)\rho^2 = -\sigma(h).$$

It is easily verified that if we rule out the "singular" solutions $\rho = \pm 1$ the equation is satisfied by $\rho = \cosh(u + k_1)$ if $h > 0$ and $\rho = \cos(u + k_2)$ if $h < 0$. According to (7.2), where the choice of T is not yet specified we are free to choose k_1 and k_2. Let them be zero. Then, by (8.1), we obtain $r = a(e \cosh u - 1)$ if $h > 0$ and $r = a(1 - e \cos u)$ if $h < 0$. According to (7.2), we have $k\, dt = r\, du$. Since $u = 0$ when $t = T$, we can integrate both sides to obtain $k(t - T) = \int_0^u r\, du$. Substituting for r each of the functions just obtained we get the parametric pairs

$$r = a(e \cosh u - 1) \qquad \text{if } h > 0,$$
$$n(t - T) = e \sinh u - u \tag{8.2}$$

and

$$r = a(1 - e \cos u) \qquad \text{if } h < 0,$$
$$n(t - T) = u - e \sin u. \tag{8.3}$$

The coefficient n is defined by $n = k/a$ or

$$n = \mu^{1/2} a^{-3/2}, \tag{8.4}$$

and is called the *mean motion*. Observe that in the case of elliptic motion $n = 2\pi/p$, where p is the period (see Ex. 5.3), so that n is simply the frequency.

Observe that if $u = 0$, then $t = T$ and $r = a|e - 1|$. It follows from the equation of the orbit, namely

$$r = \frac{a|e^2 - 1|}{1 + e \cos f}, \tag{8.5}$$

that if $c \neq 0$, T is a time of pericenter passage. On the other hand, if $c = 0$, then $e = 1$, so that $r = 0$ and T is a time of collision with or emission from the origin.

From this point on it is well to separate the cases $h > 0$ and $h < 0$. This is done in Secs. 9 and 10.

EXERCISE 8.1. Show from the Eqs. (8.2) that if $h > 0$, then as $|t| \to \infty$ the ratio $r/|t|$ approaches $\sqrt{2h}$, provided that the value $r = 0$ is not reached at a finite value of t. This gives an alternative solution of Ex. 5.5.

EXERCISE 8.2. Show from the formula $r + \mathbf{e} \cdot \mathbf{r} = c^2/\mu$ that if $h > 0$, $c \neq 0$, the unit vector $\mathbf{r}/r$ approaches a limit vector $\mathbf{l}$ as $t \to \infty$ and that $\mathbf{e} \cdot \mathbf{l} = -1$. Then, according to the formula

$$\mu(\mathbf{c} \times \mathbf{e}) = c^2 \mathbf{v} - \mu \, \frac{\mathbf{c} \times \mathbf{r}}{r} \; ,$$

easily derived from (4.2), the vector $\mathbf{v}$ also approaches a limit $\mathbf{V}$. What is the length of $\mathbf{V}$?

*EXERCISE 8.3. By matching each of Eqs. (8.2) and (8.3) with (8.5) pair-wise, obtain these formulas connecting true and eccentric anomalies:

$$\tan \frac{f}{2} = \left(\frac{e+1}{e-1} \right)^{1/2} \tanh \frac{u}{2} \; , \qquad h > 0$$

$$\tan \frac{f}{2} = \left(\frac{1+e}{1-e} \right)^{1/2} \tan \frac{u}{2} \; , \qquad h < 0.$$

*EXERCISE 8.4. Show that for each value of t each of the equations

$$n(t - T) = e \sinh u - u, \qquad e \geqslant 1$$

$$n(t - T) = u - e \sin u, \qquad 0 < e \leqslant 1$$

has a unique solution u. They are known as *Kepler's equations*.

9. POSITION ON THE ORBIT: THE CASE $h > 0$

We start with the Eqs. (8.2), which we reproduce here as

$$r = a(e \cosh u - 1) \qquad (9.1)$$

and

$$n(t - T) = e \sinh u - u. \qquad (9.2)$$

The first step is the determination of T from $\mathbf{r}_0$ and $\mathbf{v}_0$. Starting with the formulas

$$\mathbf{r} \cdot \mathbf{v} = r\dot{r} = rr'\dot{u} = rr'kr^{-1} = kr' = \sqrt{\mu a}\, e \sinh u,$$

we see that the value of u_0 of u at $t = 0$ is given by $(\mathbf{r}_0 \cdot \mathbf{v}_0) = \sqrt{\mu a}\, e \sinh u_0$. Now let $t = 0$ in (9.2) and we find that T is given by $-nT = e \sinh u_0 - u_0$. Remember that if $c = 0$, then time T corresponds to a collision or emission; hence (9.2) is valid only if $t < T$ in the first case and $t > T$ in the second.

Now to determine the location at a time t, we must solve (9.2) for u and then substitute into (9.1) to obtain the corresponding value of r. If $c = 0$ the motion is linear and the location is complete. If $c \neq 0$ there are two possible values of f which satisfy

$$r = \frac{a(e^2 - 1)}{1 + e \cos f}.$$

Clearly, we must choose $f > 0$ if $t > T$ and $f < 0$ if $t < T$.

The quantity $l = n(t - T)$ is known as the *mean anomaly*. If t is given, l is determined and the main problem in the preceding computation is the solution of $l = e \sinh u - u$ for u. A solution for the function $u = u(l)$ in some recognizable form is lacking, and the problem is usually treated as a numerical one. A simple procedure is this. For the given value of l, plot the line $y = l + u$ and the curve $y = e \sinh u$. Then their intersection yields a value u_0 which, because of the roughness of method, will generally be a first approximation to the answer.

Improved approximations can be obtained by Newton's method, as follows. Let $y = l + u - e \sinh u$. We seek the value of u for which y vanishes, starting with the approximation $u = u_0$. Draw the tangent to the curve at u_0 and find where this tangent hits the y-axis. This gives an improved value u_1 and the method can be repeated. Analytically, if u_n is the result of n successive uses of the method, then

$$u_{n+1} = u_n + \frac{l + u_n - e \sinh u_n}{e \cosh u_n - 1} . \text{*}$$

EXERCISE 9.1.   Solve the equation

$$1.667 = 2 \sinh u - u$$

numerically.

10.   POSITION ON THE ORBIT: THE CASE $h < 0$

The parametric equations in the case of negative energy read

$$r = a(1 - e \cos u), \qquad (10.1)$$

and

$$l = u - e \sin u, \qquad (10.2)$$

where l is the mean anomaly $n(t - T)$.

---

* For more about this subject consult P. Herget. *The Computation of Orbits*, privately printed, Cincinnati, Ohio, 1948.

The quantity u has an important geometric meaning if $c \neq 0$. In fact, in most treatments of the subject, u is introduced by its geometric interpretation rather than as an analytical device. The motivation for following the procedure we have adopted is the fact that in the three-body problem to be discussed later an analogue of (7.2) has important significance, whereas the geometric meaning of u will be lost.

To describe the geometry, consider the ellipse of Fig. 3, which corresponds to an orbit. The center of attraction is O, P is the pericenter, and C is the center of the ellipse. The arrow indicates the direction of motion. Let Q be a position of the particle when the true anomaly is f. Project Q to that point S of the circle for which SQ is perpendicular to CP. Then the angle PCS is u. The proof follows from (10.1) and is left to the reader.

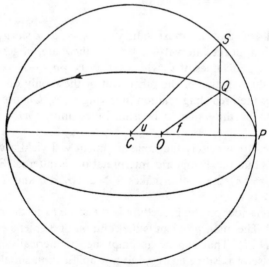

Figure 3

Observe that as Q moves around the ellipse, as indicated by the arrow, u and f each change by 2π every time Q goes through pericenter. As in the earlier cases, we must determine T. Since the particle goes through P periodically, T is not uniquely determined by $\mathbf{r}_0$, $\mathbf{v}_0$. We shall agree, however, to choose T as follows if $c \neq 0$. If at $t = 0$, $f_0 > 0$, that is, if the particle is on the upper half of the ellipse, then T is the first time before $t = 0$ that the particle went through P. On the other hand, if $f_0 < 0$, then T is the first time after $t = 0$ that the particle will go through pericenter. Analytically the computation goes this way: Since

$$\mathbf{r} \cdot \mathbf{v} = r\dot{r} = rr'\dot{u} = rr'kr^{-1} = \sqrt{\frac{\mu}{a}}\, r'$$

$$= \sqrt{\mu a}\, e \sin u, \tag{10.3}$$

it follows that u_0 must satisfy $\mathbf{r}_0 \cdot \mathbf{v}_0 = \sqrt{\mu a}\, e \sin u_0$. In addition, in the interval $-\pi < u \leqslant \pi$ there are, in general, exactly two values of u_0 which satisfy $r_0 = a(1 - e \cos u_0)$, each the negative of the other. But of these only one can satisfy the preceding relation involving $\mathbf{r}_0 \cdot \mathbf{v}_0$. Choose that one to be the value to be substituted into $-nT = u_0 - e \sin u_0$.

If $c = 0$, precisely the same argument will yield a value of T, but the geometric interpretation is altered. Since $\mathbf{r}_0 \cdot \mathbf{v}_0 = r_0 \dot{r}_0$, the choice makes $T > 0$ if $\dot{r}_0 < 0$ and $T < 0$ if $\dot{r}_0 > 0$.

From now on the procedure is the same as in the cases $h \geqslant 0$. The main problem is the solution of Kepler's equation (10.2). That can be accomplished numerically as in the case of positive energy, but a simplification should be observed. The equation is unchanged if we simultaneously add or subtract any multiple of 2π to both l and u. Therefore, when l is given, add or subtract a multiple of 2π to bring it into the range $-\pi \leqslant l \leqslant \pi$. Moreover, the equation is unchanged if l and u are simultaneously replaced by $-l$ and $-u$, respectively. This means that u is an odd function of l and it is enough to solve the equation when $0 \leqslant l \leqslant \pi$. When $l = 0$, $u = 0$ and when $l = \pi$, $u = \pi$. Therefore, the problem is reduced to the range $0 < l < \pi$. It is clear from the graph of l against u (see Fig. 4) that the values of u also lie in the range $0 < u < \pi$.

If the eccentricity is in the range $0 < e < 1$, there exist analytic solutions of the problem. We defer the discussion to Sec. 12.

*EXERCISE 10.1   Prove that if $0 < e < 1$, the function $u(l)$ defined by (10.2) has the property that $u(l) - l$ is

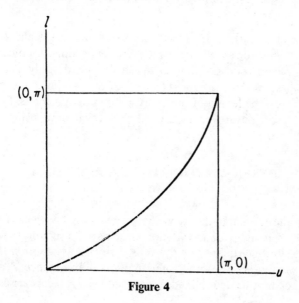

Figure 4

periodic of period 2π in l, is odd, vanishes at $l = 0$, $l = \pi$ and has a continuous derivative. Therefore, it may be expanded in a uniformly convergent Fourier series

$$u(l) - l = \sum_{n=1}^{\infty} u_n \sin nl.$$

Prove that

$$u_n = \frac{2}{\pi n} \int_0^{\pi} \cos n(u - e \sin u)du.$$

*EXERCISE 10.2. Let Q_0, Q_1 be two positions on an elliptic orbit, and let u_0, u_1 be the corresponding eccentric anomalies. Assume $u_1 > u_0$. Prove that the distance Q_0Q_1 is $2a \sin \alpha \sin \beta$, where $\alpha = \frac{1}{2}(u_1 - u_0)$ and β is defined by $\cos \beta = e \cos \frac{1}{2}(u_1 + u_0)$, $0 < \beta < \pi$.

EXERCISE 10.3. Prove Lambert's theorem, which says that for an elliptic orbit the time occupied in moving from one position to another depends only on the sum of the distances from O of the two positions, and on the length of the chord joining the positions. (This will be proved in Sec. 11, but try it now, using Ex. 10.2.)

11. DETERMINATION OF THE PATH OF A PARTICLE

In the preceding theory we have solved the problem of the determination of the motion of a particle moving under the inverse square law $f(r) = \mu r^{-2}$ on the assumption that $\mathbf{r}_0$ and $\mathbf{v}_0$ are known at some time $t = 0$. In practice, $\mathbf{r}_0$ and $\mathbf{v}_0$ cannot be determined directly, so the problems arises of the determination of the motion when other types of data are given. We shall be content with one example, highly idealized for the sake of illustration. The realistic problems are treated definitively in Herget's book mentioned at the end of Sec. 9.

Suppose the center of attraction is the center of the earth, regarded as a point mass, and that the particle is an artificial satellite moving in elliptic motion. Its positions r_0 and r_1 are observed in succession at times τ units apart. It will be assumed that the angle g swept out by the radius vector $\mathbf{r}$ in moving from $\mathbf{r}_0$ to $\mathbf{r}_1$ is small enough so that the area caught between the chord joining the observed positions and the orbit itself does not contain O. It may, however, contain the "empty" focus F, that is, the focus which is not the center of attraction. This is illustrated in Fig. 5 by the shaded regions.

The plane of motion is determined by $\mathbf{r}_0$ and $\mathbf{r}_1$. Let $\mathbf{e}$ be the (unknown) eccentric axis and f the true anomaly measured from $\mathbf{e}$. Then the conic has the equation

$$r = \frac{a(1 - e^2)}{1 + e \cos f}.$$

Suppose now that a has been found by some means. We shall show how to find the remaining constants. Let f_0 be the true anomaly of the first position. Then $f_0 + g$ is the anomaly of the second. Hence, we have the relations

$$r_1 = \frac{a(1 - e^2)}{1 + e \cos (f_0 + g)},$$

$$r_0 = \frac{a(1 - e^2)}{1 + e \cos f_0}. \tag{11.1}$$

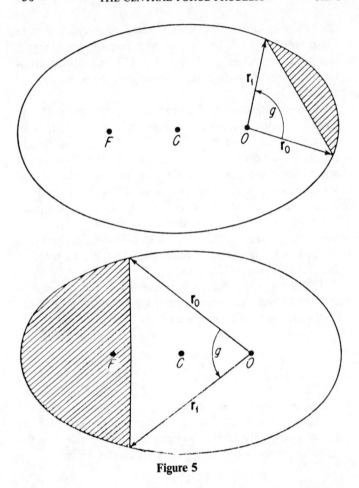

Figure 5

From these, the unknowns e and f_0 can be determined. This locates the eccentric axis, which is forward of $\mathbf{r}_0$ by the angle $-f_0$ if $f_0 < 0$, and back of it by f_0 if $f_0 > 0$. The orbit is now completely determined.

However, position on the orbit is not. For this we need to know $\mathbf{v}_0$, the velocity vector corresponding to $\mathbf{r}_0$. For then the problem becomes the initial condition problem discussed in the earlier sections. Now the components of $\mathbf{v}_0$ are $\dot{r}_0$ in the direction of $\mathbf{r}_0$, and c/r_0 perpendicular to it in the direction of motion (see Ex. 3.3). So all we need are the values of $\dot{r}_0$ and c. The latter can be found from $c^2 = \mu a (1 - e^2)$, the former from

$$\dot{r}_0^2 + \frac{c^2}{r_0^2} = \mu \left(\frac{2}{r_0} - \frac{1}{a} \right),$$

where $\dot{r}_0 > 0$ if $f_0 > 0$ and $\dot{r}_0 < 0$ if $f_0 < 0$.

There remains the determination of a whose value was assumed to be known in the preceding discussion. If the time τ between observations were a period, then a could be found from Kepler's third law. But τ is less than a period and another method must be found. The key is Lambert's theorem anticipated in Ex. 10.3.

Let u_0, u_1 be the eccentric anomalies at the two positions, where $-\pi < u_0 \leqslant \pi$, $-\pi < u_1 \leqslant \pi$. Then $r_1 = a(1 - e \cos u_1)$, $r_0 = a(1 - e \cos u_0)$ and

$$r_1 + r_0 = 2a \left[1 - e \cos \tfrac{1}{2}(u_1 - u_0) \cos \tfrac{1}{2}(u_1 + u_0) \right].$$

Therefore, using the notation of Ex. 10.2, $r_1 + r_0 = 2a(1 - \cos \alpha \cos \beta)$. Moreover, the distance ρ between the positions is given by $\rho = 2a \sin \alpha \sin \beta$. Therefore

$$r_1 + r_0 + \rho = 2a \left[1 - \cos(\alpha + \beta) \right] = 4a \sin^2 \tfrac{1}{2}(\alpha + \beta),$$

$$r_1 + r_0 - \rho = 2a \left[1 - \cos(\alpha - \beta) \right] = 4a \sin^2 \tfrac{1}{2}(\alpha - \beta).$$

Since $n(t - T) = u - e \sin u$ gives the eccentric anomaly at time t, it follows that the elapsed time τ

between observations is given by

$$n\tau = (u_1 - u_0) - e(\sin u_1 - \sin u_0)$$

$$= (u_1 - u_0) - 2e \sin \tfrac{1}{2}(u_1 - u_0)\cos \tfrac{1}{2}(u_1 + u_0)$$

$$= 2\alpha - 2 \sin \alpha \cos \beta.$$

Observe that ρ is known because r_1, r_0 and the angle g between the position vectors is known. In fact, by the cosine law $\rho^2 = r_1^2 - 2r_0 r_1 \cos g + r_0^2$. In summary, let $\epsilon = \alpha + \beta$, $\delta = \beta - \alpha$, and replace n by its value $\mu^{1/2}a^{-3/2}$. Then we have three equations

$$4a \sin^2 \frac{\epsilon}{2} = r_1 + r_0 + \rho,$$

$$4a \sin^2 \frac{\delta}{2} = r_1 + r_0 + \rho,$$

$$\mu^{1/2}\tau = a^{3/2}\big[\epsilon - \delta - (\sin \epsilon - \sin \delta)\big],$$

for the unknowns ϵ, δ, a. If ϵ and δ can be found from the first two, their values can be substituted into the third, giving one equation for the determination of a.

There is a difficulty here because the solutions for ϵ and δ are not unique. Since $-\pi < u_0 \leqslant \pi$, $-\pi < u_1 \leqslant \pi$ and $u_0 < u_1$, we know that $0 < \alpha \leqslant \pi$. Also, $0 < \beta < \pi$ by its definition. Therefore $0 < \epsilon < 2\pi$. Similarly, $-\pi < \delta < \pi$. Hence, if (ϵ_1, δ_1) is the smallest pair of positive angles satisfying the equations for ϵ and δ, the remaining pairs are $(2\pi - \epsilon_1, \delta_1)$, $(\epsilon_1, -\delta_1)$ and $(2\pi - \epsilon_1, -\delta_1)$. It turns out that the last two cases are excluded by our assumption that the shaded areas of Fig. 5 do not contain O. This is discussed by H. C. Plummer.[*]

* An Introductory Treatise on Dynamical Astronomy, New York: Dover Publications, 1960, pp. 51–52.

He shows also that the proper choice of ϵ is ϵ_1 if the shaded area does not contain F, otherwise it is $2\pi - \epsilon_1$. Therefore the equation for a is

$$\mu^{1/2}\tau = a^{3/2}\big[\epsilon_1 - \delta_1 - (\sin \epsilon_1 - \sin \delta_1)\big]$$

in the first case, and

$$\mu^{1/2}\tau = a^{3/2}\big[2\pi - \epsilon_1 - \delta_1 + (\sin \epsilon_1 + \sin \delta_1)\big]$$

in the second.

It follows, therefore, that under the given conditions, two orbits satisfy the given data.

EXERCISE 11.1. Show how Eqs. (11.1) determine e and f_0.

12. EXPANSIONS IN ELLIPTIC MOTION

We have already seen in Ex. 10.1 that in case $0 < e < 1$ Kepler's equation

$$l = u - e \sin u \qquad (12.1)$$

has a solution which permits expansion of $u(l) - l$ in a uniformly convergent sine series

$$u(l) - l = \sum_{n=1}^{\infty} u_n \sin nl. \qquad (12.2)$$

According to the standard formula for the coefficients of a sine series,

$$u_n = \frac{2}{\pi} \int_0^{\pi} \big[u(l) - l\big]\sin nl \; dl.$$

To evaluate the integral, write this as

$$u_n = -\frac{2}{\pi n} \int_0^{\pi} \big[u(l) - l\big]d \cos nl$$

and integrate by parts to obtain

$$u_n = \frac{2}{\pi n} \int_0^\pi \cos nl \, d[u(l) - l]$$

$$= \frac{2}{\pi n} \int_0^\pi \cos nl \, du(l) - \frac{2}{\pi n} \int_0^\pi \cos nl \, dl$$

$$= \frac{2}{\pi n} \int_0^\pi \cos nl \, du(l).$$

Now let $l = u - e \sin u$, according to (12.1). The limits of integration are unchanged, so that

$$u_n = \frac{2}{\pi n} \int_0^\pi \cos n(u - e \sin u) du.$$

The Bessel functions $J_n(x)$ are well-known in many parts of mathematics and can be defined in a variety of equivalent ways. For our purpose this one is best:

$$J_n(x) = \frac{1}{\pi} \int_0^\pi \cos(nu - x \sin u) du.$$

It follows that $u_n = (2/n) J_n(ne)$ and Eq. (12.2) takes the form

$$u = l + 2 \sum_{n=1}^\infty n^{-1} J_n(ne) \sin nl,$$

so that by (12.1) once again,

$$e \sin u = 2 \sum_{n=1}^\infty n^{-1} J_n(ne) \sin nl.$$

These expansions have many important consequences, including formulas for the position of the particle. A rigorous treatment is given by A. Wintner.*

Here we give only one formal consequence of the preceding theorem.

According to (12.1), $dl/du = 1 - e \cos u = r/a$. Therefore, if we differentiate the last series with respect to l we obtain

$$(e \cos u)\frac{a}{r} = 2 \sum_{n=1}^{\infty} J_n(ne)\cos nl.$$

Since $e \cos u = 1 - (r/a)$,

$$\frac{a}{r} = 1 + 2 \sum_{n=1}^{\infty} J_n(ne)\cos nl.$$

EXERCISE 12.1. Give a proof of the last formula starting with

$$(1 - e \cos u)^{-1} = \frac{c_0}{2} + \sum_{n=1}^{\infty} c_n \cos nl,$$

where

$$c_n = \frac{2}{\pi} \int_0^{\pi} (1 - e \cos u)^{-1} \cos nl \, dl$$

$$= \frac{2}{\pi} \int_0^{\pi} \cos nl \, du.$$

---

* *The Analytical Foundations of Celestial Mechanics*, Princeton University Press, 1947, pp. 204–22.

13.   ELEMENTS OF AN ORBIT

In the preceding treatment of the non-linear case $c \neq 0$, the coordinate system used is indeterminate in one respect. In the plane of motion perpendicular to $\mathbf{c}$ (see Figs. 1 and 2), a system of axes x, y is installed to form a right-handed coordinate system with respect to $\mathbf{c}$. Since $r^2\dot{\theta} = c$, the motion is in the direction of increasing θ. The orbit is completely determined by c, e and position on it by T, time of pericenter passage. Alternatively, we may say that *once the x-axis is in place* everything is determined by the quantities

$$e, \begin{cases} a & \text{if} \quad e \neq 1, \\ c & \text{if} \quad e = 1, \end{cases} \omega, T. \qquad (13.1)$$

Now suppose, as is the case in practice, that a *prescribed* coordinate system X, Y, Z is given with its origin at O. The problem is now that of describing the motion in the prescribed system. Such a system is illustrated in Fig. 6, along with the position of the vector $\mathbf{c}$. What must be done is to find a unique prescription for the x-axis. Then points in the x, y, $\mathbf{c}$ coordinate system can be described in the X, Y, Z system.

If $\mathbf{c}$ falls on the positive Z-axis, it is reasonable to choose the x-axis of Fig. 2 to fall along X; and if $\mathbf{c}$ falls on the negative Z-axis, it is reasonable to choose the x-axis to fall along Y in order to preserve the right-handed orientation.

Otherwise the plane of motion is determined by i, the angle from Z to $\mathbf{c}$, and by the line of intersection of that plane with the XY-plane. The angle i is called the *angle of inclination*, or simply the inclination, and the line, shown dotted in Fig. 6, the *line of nodes*.

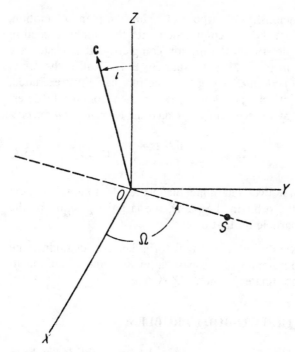

Figure 6

It is now customary to choose the x-axis in the plane of
motion as follows. First exclude that rare case of non-
elliptic motion in which the dotted line falls along the axis
of the conic. Then the orbit will cut twice through the line
of nodes, once on its way "up," the other on the way
down. Let S be the point at which the particle cuts on its
upward journey. S is called the *ascending* node, and OS is
chosen as the positive x-axis. The angle XOS, measured
counterclockwise as seen from the positive Z-axis, is
called the *longitude of the ascending node*. The angles i and

Ω accomplish the purpose of fixing the plane of motion. Therefore they, in conjunction with the numbers listed in (13.1), determine the motion completely. It is customary to use in place of ω the sum $\varpi = \Omega + \omega$, called the *longitude of pericenter*. Except for the rare cases just excluded, the orbit and position on it are then completely determined by the six numbers, called the *elements of the orbit*:

$$i, \Omega; \; e, \; \begin{Bmatrix} a & \text{if} & e \neq 1, \\ c & \text{if} & e = 1, \end{Bmatrix} \varpi; \; T. \qquad (13.2)$$

The first two determine the plane of motion, the next three the orbit in the plane, the last the position of the mass particle on that orbit.

> EXERCISE 13.1. Find formulas for changing the coordinates of the particle in its plane of motion to coordinates in the XYZ system.

14. THE TWO-BODY PROBLEM

Once the solution of the central force problem has been achieved, it is possible to solve what appears at first sight to be a more complicated problem: to describe the motion of a system of *two* mass particles moving according to their mutual gravitational attraction. This is known as the *two-body problem*, although the name *two-particle problem* would be a more accurate description.*

Let O represent a fixed point in the space of motion (see Fig. 7), let m_1, m_2 denote the masses of the two particles, $\mathbf{r}_1$, $\mathbf{r}_2$ their positions, and r the distance between them. Clearly, $r = |\mathbf{r}_2 - \mathbf{r}_1|$. According to Newton's law of

---

* The two-body problem for finite bodies is unsolved.

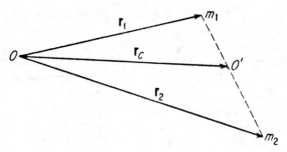

Figure 7

universal gravitation, the force of attraction between the particles is $Gm_1 m_2 r^{-2}$, where G is a constant depending solely on the choice of units. The differential equations are then

$$m_1 \ddot{\mathbf{r}}_1 = \frac{Gm_1 m_2}{r^2} \, \frac{\mathbf{r}_2 - \mathbf{r}_1}{r} \, ,$$

$$m_2 \ddot{\mathbf{r}}_2 = \frac{Gm_2 m_1}{r^2} \, \frac{\mathbf{r}_1 - \mathbf{r}_2}{r} \, ,$$

(14.1)

and it is assumed that initial values of $\mathbf{r}_1, \mathbf{r}_2, \dot{\mathbf{r}}_1, \dot{\mathbf{r}}_2$ are given.

It is possible to reduce the problem to the central force problem by the following procedure, called the *reduction to relative coordinates*. Divide the first of Eqs. (14.1) by m_1, the second by m_2 and subtract the first from the second. If $\mathbf{r}$ denotes $\mathbf{r}_2 - \mathbf{r}_1$ we find that

$$\ddot{\mathbf{r}} = -\mu r^{-3}\mathbf{r}, \qquad \mu = G(m_1 + m_2). \qquad (14.2)$$

Clearly, initial values of $\mathbf{r}$ and $\mathbf{v}$ are known from the corresponding values for the original system (14.1). But (14.2) is precisely *the central force problem with a special choice of* μ, and all the preceding theory is applicable. Once **r** is determined, so is the right-hand side of each Eq. (14.1), from which both $\mathbf{r}_1$ and $\mathbf{r}_2$ can be obtained. In summary, *each* particle moves as if it were a unit mass attracted to a fixed center located at the other mass, with $\mu = G(m_1 + m_2)$. The orbit of each, as seen from the other, is called a *relative orbit*. Equation (14.2) is unchanged if **r** is replaced by $-\mathbf{r}$. Therefore, the relative orbits are geometrically identical.

Another procedure, called *the reduction to barycentric coordinates*, is also important. First add Eqs. (14.1) together as they stand. Then $m_1\ddot{\mathbf{r}}_1 + m_2\ddot{\mathbf{r}}_2 = 0$. This has an important interpretation. Let

$$\mathbf{r}_c = \frac{m_1\mathbf{r}_1 + m_2\mathbf{r}_2}{m_1 + m_2}$$

denote the position vector of the center of mass O' of the two particles. Clearly it lies on the line joining them (see Fig. 7). Then $\ddot{\mathbf{r}}_c = 0$. It follows that

$$\mathbf{r}_c = \mathbf{a}t + \mathbf{b}, \tag{14.3}$$

where **a** and **b** are constant vectors determined by the initial conditions. This gives the principle of *conservation of linear momentum*: the center of mass moves in a straight line with uniform velocity. The system (14.1) is of order twelve (two vector equations make six scalar equations, and each is of the second order). The vectors **a** and **b** provide six constants of the motion, which leaves six more to be accounted for.

To discover the other six, we move the origin of coordinates to the center of mass. This means that in (14.1) we replace $\mathbf{r}_1$ by $\mathbf{r}_1 - \mathbf{r}_c$ and $\mathbf{r}_2$ by $\mathbf{r}_2 - \mathbf{r}_c$. Since $\ddot{\mathbf{r}}_c = 0$, the Eqs. (14.1) remain unaltered by the change, and we may suppose from this point on that the origin is *fixed* at O', the center of mass. O' itself moves according to (14.3) and we are now studying the motion of m_1 and m_2 relative to O', which we now rename O. $\mathbf{r}_1$ and $\mathbf{r}_2$ are positions relative to the center of mass.

We now proceed in this way. Let r_1 and r_2 denote the lengths of $\mathbf{r}_1$ and $\mathbf{r}_2$, respectively. Then

$$r = r_1 + r_2, \qquad m_1 r_1 = m_2 r_2, \qquad m_1 \mathbf{r}_1 + m_2 \mathbf{r}_2 = 0. \quad (14.4)$$

This enables us to rewrite (14.1) as a pair of equations which are formally independent of one another, namely:

$$\begin{aligned}
\ddot{\mathbf{r}}_1 &= -\left(G m_2^3 M^{-2}\right) r_1^{-3} \mathbf{r}_1, \\
\ddot{\mathbf{r}}_2 &= -\left(G m_1^3 M^{-2}\right) r_2^{-3} \mathbf{r}_2.
\end{aligned} \quad (14.5)$$

Actually *one* of these suffices since $m_1 \mathbf{r}_1 + m_2 \mathbf{r}_2 = 0$. Since each is of the form (1.1), with a special value of μ, we have accounted for six more constants, namely the elements of either orbit relative to the center of mass.

The conclusion is that the *center of mass moves uniformly and each of the particles moves with respect to that center of mass as if a fictitious force of attraction were located there with* $\mu = G m_2^3 M^{-2}$ for the first mass, $\mu = G m_2^3 M^{-2}$ for the *second*.

In what follows, we suppose the origin fixed at the center of mass. The *potential energy* of the system is defined to be $-U^*$, where

$$U^* = G m_1 m_2 r^{-1}, \quad (14.6)$$

and the *kinetic energy* T^* is defined to be

$$T^* = \tfrac{1}{2}\left(m_1 v_1^2 + m_2 v_2^2\right), \quad (14.7)$$

where $\mathbf{v}_1 = \dot{\mathbf{r}}_1$ and $\mathbf{v}_2 = \dot{\mathbf{r}}_2$. Now let us examine each of the Eqs. (14.5) as if it corresponds to a central force problem. According to (3.1), each corresponds to a constant total "energy" defined, respectively, by

$$h_1 = \tfrac{1}{2} m_1 v_1^2 - G m_1 m_2^3 M^{-2} r_1^{-1} \equiv T_1 - U_1$$

and

$$h_2 = \tfrac{1}{2} m_2 v_2^2 - G m_2 m_1^3 M^{-2} r_2^{-1} \equiv T_2 - U_2.$$

Using (14.4), we can conclude that

$$T^* = T_1 + T_2, \qquad U^* = U_1 + U_2.$$

Moreover,

$$\frac{h_1}{h_2} = \frac{U_1}{U_2} = \frac{T_1}{T_2} = \frac{m_2}{m_1}.$$

Therefore the various energies (kinetic, potential, total) are split between the masses m_1 and m_2 in the ratio m_2/m_1.

EXERCISE 14.1. The shape of an orbit in the central force problem is determined by the sign of h. Prove from this that in the two-body problem the orbit of each mass, relative to the center of mass, is the same kind of conic for each.

*EXERCISE 14.2. Starting with Eq. (14.2) for the relative motion of two particles, study the behavior of r at an instant of collision. Notice that (7.1) applies with $c = 0$, $\mu = G(m_1 + m_2)$, so that $r\dot{r}^2 = 2(\mu + hr)$. Since $r \to 0$ at a collision we have $r\dot{r}^2 \to 2\mu$ when $t \to t_1$, the time of collision. This is independent of the sign of h. Conclude that $|\dot{r}| r^{1/2} \to \sqrt{2\mu}$ and hence that $r|t - t_1|^{-2/3} \to (9/2\mu)^{1/3}$ as $t \to t_1$.

15.   THE SOLAR SYSTEM

The real solar system is very complicated. Mainly for the purpose of illustrating the preceding theory, we describe a simplified solar system. It consists of ten particles, one of which, the *sun*, carries most of the total mass. The other nine are *planets*. Since most of the mass is in the sun, it will be supposed that each of the planets moves independently of the others and is acted on only by the sun. The result is that we have nine independent two-body systems each consisting of the sun and one planet. Motion will be discussed relative to the sun, in accordance with the first part of Sec. 14. Then each planet is governed by Eq. (14.2), with $\mu = G(m_s + m_p)$, m_s being the mass of the sun and m_p that of the planet. Consistent with this, each planet moves in an ellipse with the sun at one focus. Let n_p and a_p denote the mean motion and the semi-major axis, respectively. Then, according to Kepler's third law (8.8) $n_p^2 a_p^3 = G(m_s + m_p)$. It follows that for two distinct planets p and q we have the law

$$\frac{n_p^2 a_p^3}{n_q^2 a_q^2} = \frac{1 + m_p/m_s}{1 + m_q/m_s} . \tag{15.1}$$

Since m_s is very large compared to m_p and m_q, the ratio on the right-hand side is very close to 1. Therefore, $n_p^2 a_p^3$ is almost (but not quite) the same for each of the planets. This is the original form of Kepler's third law.

To describe the actual orbits of the planet, it is customary to list the elements relative to the following coordinate system. (See Fig. 8, which is a special example of Fig. 6.) The origin is taken as the sun, the plane of the earth's orbit is the XY-plane. This orbit is known as the *ecliptic*, the XY-plane as the *plane of the ecliptic*. The

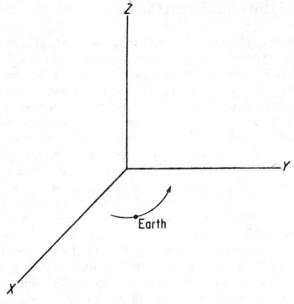

Figure 8

X-axis is directed towards a point among the stars known as the *vernal equinox*. A precise definition can be found in the textbooks on astronomy. All that matters for our purpose is that it is to be regarded as fixed. Each orbit is then defined by its elements i, Ω, which give the plane of motion; a, e, ϖ, which describe the conic in that plane; and position on the orbit can be found from T, the date of perihelion passage.

We append a table of the elements of the nine major planets. In addition, we include the period p (measured in days) and the mass M (relative to the earth, which is taken to be of mass 1). Distance is measured in astronomical units, where one unit is the length of the

Table of Elements, 1900

	ι	Ω	a	e	ϖ	T	ρ	m
Mercury	7°.00	47°.14	.387	.206	75°.90	Mar. 3, 1900	87.97	.053
Venus	3°.59	75°.78	.723	.007	130°.15	Apr. 1, 1900	224.7	.815
Earth	0°.00	0°.00	1.000	.017	101°.22	Jan. 1, 1900	365.26	1.000
Mars	1°.85	48°.78	1.524	.093	334°.22	Mar. 18, 1900	686.98	.107
Jupiter	1°.31	99°.44	5.203	.048	12°.72	June 1, 1904	4,332.6	318.00
Saturn	2°.5	112°.79	9.546	.056	91°.09	Feb. 20, 1915	10,759.	95.22
Uranus	0°.77	73°.48	19.20	.047	169°.05	May 20, 1966	30,687.	14.55
Neptune	1°.78	130°.68	30.09	.009	43°.83	Sept. 15, 2042	60,184.	17.23
Pluto	17°.14	108°.95	39.5	.247	222°.8	Aug. 5, 1989	90,700.	.9(?)

semi-major axis of the earth. Time of perihelion passage T is the first date of this event after December 31, 1899. Angles are given in degrees.

16. DISTURBED MOTION

We return to the problem of central attraction according to the inverse square law. The governing differential equation is $\ddot{\mathbf{r}} = -\mu r^{-3}\mathbf{r}$. Suppose that in addition to the central force, the moving particle is subjected to an additional force. This may be due to the attraction of some other body, to air resistance, or any other cause. The equation becomes

$$\dot{\mathbf{v}} = \ddot{\mathbf{r}} = -\mu r^{-3}\mathbf{r} + \mathbf{F}, \qquad (16.1)$$

where $\mathbf{F}$ is the force per unit mass. We shall call the motion subject to the extra force *disturbed*, and the motion with $\mathbf{F} = 0$ *undisturbed*.

Suppose that the particle is moving subject to the disturbing force which at some instant t is suddenly wiped out. Let $\mathbf{r}(t)$, $\mathbf{v}(t)$ represent the position and velocity at that instant. From then on the particle will move according to the theory described earlier in the chapter. In particular, we can define the vectors $\mathbf{c}$, $\mathbf{e}$ and the time of pericenter passage T just as before, regarding $\mathbf{r}(t)$ and $\mathbf{v}(t)$ as the initial data. But $\mathbf{c}$ and $\mathbf{e}$ are dependent on the instant t at which $\mathbf{F}$ is wiped out. They are, therefore, functions of t.

At each instant t during the disturbed motion we can look at the particle in two ways: it is moving on its real orbit, or it is about to move on its undisturbed orbit, called the *osculating* orbit. With this as the clue, we are going to study the real orbit by finding how the undisturbed orbit changes with time. In other words, we shall see how $\mathbf{c}$, $\mathbf{e}$ and T change with time. Since at each instant of time these quantities determine the elements of the undisturbed orbit, this will enable us to find how the elements of the undisturbed orbit change with time.

We shall start with the definition $\mathbf{c} = \mathbf{r} \times \mathbf{v}$, where $\mathbf{r}$ and $\mathbf{v}$ are the position and velocity on the disturbed orbit, so that $\mathbf{c}$ depends on t. Then $\dot{\mathbf{c}} = \mathbf{r} \times \dot{\mathbf{v}}$, or by (16.1),

$$\dot{\mathbf{c}} = \mathbf{r} \times (-\mu r^{-3}\mathbf{r} + \mathbf{F}) = \mathbf{r} \times \mathbf{F}, \qquad (16.2)$$

since $\mathbf{r} \times \mathbf{r} = 0$.

We define the vector $\mathbf{e}$ by the equation

$$\mu\left(\frac{\mathbf{r}}{r} + \mathbf{e}\right) = \mathbf{v} \times \mathbf{c}. \qquad (16.3)$$

Since $\mathbf{e}$ is a function of time we can conclude that

$$\mu\left(\frac{d}{dt}\frac{\mathbf{r}}{r} + \dot{\mathbf{e}}\right) = \dot{\mathbf{v}} \times \mathbf{c} + \mathbf{v} \times \dot{\mathbf{c}}.$$

Now replace $\dot{\mathbf{v}}$ according to (16.1), $\dot{\mathbf{c}}$ according to (16.2) and $(d/dt)(\mathbf{r}/r)$ according to (2.3). Then

$$\mu\dot{\mathbf{e}} = \mathbf{F} \times \mathbf{c} + \mathbf{v} \times (\mathbf{r} \times \mathbf{F}). \qquad (16.4)$$

Let t be an instant of time at which $c \neq 0$ and $e \neq 0$ and let f be the angle from $\mathbf{e}$ to $\mathbf{r}$. Then f is the true anomaly of the particle regarded as being on its undisturbed orbit at that instant.

We introduce a coordinate system at the instant t. Its origin is O and the axes are $\mathbf{c}$, $\mathbf{r}$ and $\boldsymbol{\alpha}$ where $\boldsymbol{\alpha}$ is defined by $\boldsymbol{\alpha} = \mathbf{c} \times \mathbf{r}$. (See Fig. 9.) Clearly

$$\boldsymbol{\alpha} = \mathbf{c} \times \mathbf{r}, \quad r^2\mathbf{c} = \mathbf{r} \times \boldsymbol{\alpha}, \quad c^2\mathbf{r} = \boldsymbol{\alpha} \times \mathbf{c}. \qquad (16.5)$$

The vector $\mathbf{v}$ lies in the plane perpendicular to $\mathbf{c}$, so that

$$\mathbf{v} = A\mathbf{r} + B\boldsymbol{\alpha}. \qquad (16.6)$$

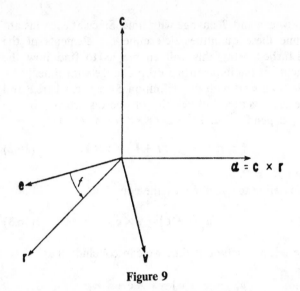

Figure 9

We proceed to compute A and B. We have $\mathbf{c} = \mathbf{r} \times \mathbf{v}$ $= \mathbf{r} \times (A\mathbf{r} + B\boldsymbol{\alpha}) = A(\mathbf{r} \times \mathbf{r}) + B(\mathbf{r} \times \boldsymbol{\alpha})$, so that, by (16.5), $\mathbf{c} = B\mathbf{r}^2\mathbf{c}$ or $B = 1/r^2$. Also by (16.6), $\mathbf{r} \cdot \mathbf{v} = A\mathbf{r} \cdot \mathbf{r} = Ar^2$, since $\boldsymbol{\alpha} \cdot \mathbf{r} = 0$. To finish the calculation of A we note that $\mathbf{e} \cdot \boldsymbol{\alpha} = e\alpha \cos(f + 90°)$. Also $\alpha = cr$. Therefore $\mathbf{e} \cdot \boldsymbol{\alpha} = - ecr \sin f$. Since $\mathbf{e} \cdot \mathbf{r} = er \cos f$, it follows on taking the dot product of both sides of (16.6) with $\mathbf{e}$ that $\mathbf{e} \cdot \mathbf{v} = Aer \cos f - Becr \sin f$. But according to (16.3), $\mathbf{r} \cdot \mathbf{v} + r(\mathbf{e} \cdot \mathbf{v}) = 0$. Therefore,

$$Ar^2 = \mathbf{r} \cdot \mathbf{v} = -r(\mathbf{e} \cdot \mathbf{v})$$

$$= -Aer^2 \cos f + Becr^2 \sin f.$$

But $Br^2 = 1$. It follows that $Ar^2(1 + e \cos f) = ec \sin f$. Since, at the instant t, $r = (c^2/\mu)(1 + e \cos f)^{-1}$, we get $A = \mu e r^{-1} c^{-1} \sin f$.

Substitute from (16.6) into (16.4) to get rid of $\mathbf{v}$. Using the fact that $Br^2 = 1$ and expanding the triple products, we find that

$$\mu\dot{\mathbf{e}} = \mathbf{F} \times \mathbf{c} - Ar^2\mathbf{F} + \left[A(\mathbf{F} \cdot \mathbf{r}) + r^{-2}(\mathbf{F} \cdot \boldsymbol{\alpha}) \right]\mathbf{r}. \quad (16.7)$$

We interrupt with an exercise.

*EXERCISE 16.1.   Write $\mathbf{F}$ in terms of its components F_c, F_r, F_α in the direction of the coordinate axes, that is, $\mathbf{F} = F_c c^{-1}\mathbf{c} + F_r r^{-1}\mathbf{r} + F_\alpha \alpha^{-1}\ \boldsymbol{\alpha}$. Show that the basic equations (16.2) and (16.7) become, respectively,

$$\dot{\mathbf{c}} = rc^{-1}F_\alpha\mathbf{c} - c^{-1}F_c\boldsymbol{\alpha} \quad (16.8)$$

and

$$\mu\dot{\mathbf{e}} = 2cr^{-1}F_\alpha\mathbf{r} - \left(r^{-1}F_r + Arc^{-1}F_\alpha\right)\boldsymbol{\alpha} - Ar^2c^{-1}F_c\mathbf{c},$$

$$(16.9)$$

where, as before, $A = \mu e r^{-1} c^{-1} \sin f$.

Dot multiply both sides of (16.8) by $\mathbf{c}$ to obtain

$$\dot{c} = rF_\alpha. \quad (16.10)$$

17.   DISTURBED MOTION: VARIATION OF THE ELEMENTS

Now let X, Y, Z be a coordinate system, as described in Sec. 13. We wish to determine how the disturbed motion looks in this coordinate system. At each instant of time we shall regard the particle as being on its undisturbed orbit with the associated constants i, Ω, ω, e, c, T and ask how these vary with the time as the particle moves through its successive undisturbed orbits.

We already know from (16.10) that

$$\dot{c} = rF_\alpha. \qquad (17.1)$$

Now let **i**, **j**, **k** denote unit vectors in the X, Y, Z directions (see Fig. 10) and let the line of nodes be directed along **n**, where $\mathbf{n} = \mathbf{k} \times \mathbf{c}$.* Clearly, $n = kc \sin i = c \sin i$. Also, because $\boldsymbol{\alpha} = \mathbf{c} \times \mathbf{r}$ we know that $\mathbf{k} \cdot \boldsymbol{\alpha} = \mathbf{k} \cdot (\mathbf{c} \times \mathbf{r}) = \mathbf{k} \times \mathbf{c} \cdot \mathbf{r} = \mathbf{n} \cdot \mathbf{r} = nr \cos(\omega + f) = cr \sin i \cos(\omega + f)$.

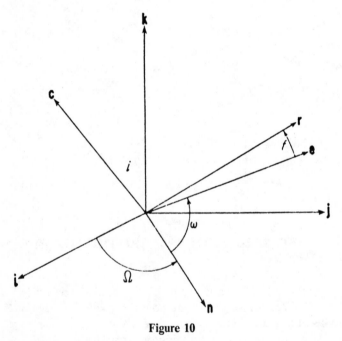

Figure 10

---

* This use of **n** is confined to the present section.

We start again with $\mathbf{c} \cdot \mathbf{k} = ck \cos i = c \cos i$, so that, according to (17.1),

$$\dot{\mathbf{c}} \cdot \mathbf{k} = rF_\alpha \cos i - c \sin i \frac{di}{dt} .$$

From (16.8) and our computation of $\mathbf{k} \cdot \boldsymbol{\alpha}$, we get

$$\dot{\mathbf{c}} \cdot \mathbf{k} = rc^{-1}F_\alpha c \cos i - c^{-1}F_c cr \sin i \cos(\omega + f).$$

From the last two equations it follows that

$$\frac{di}{dt} = rc^{-1}F_c \cos(\omega + f). \qquad (17.2)$$

We turn to the computation of $\dot{e}$. According to Fig. 9, we know that $\boldsymbol{\alpha} \cdot \mathbf{e} = \alpha e \cos(f + 90°) = -rce \sin f$. Now dot multiply both sides of (16.9) by $\mathbf{e}$ to obtain

$$\mu e \dot{e} = 2cr^{-1}F_\alpha \left(\frac{c^2}{\mu} - r \right) - \left(r^{-1}F_r + Arc^{-1}F_\alpha \right)(-rce \sin f)$$

$$= ceF_r \sin f + ce(1 + e \cos f)^{-1}F_\alpha(e + 2 \cos f + e \cos^2 f),$$

or

$$\mu c^{-1}\dot{e} = F_r \sin f + F_\alpha(e + 2 \cos f + e \cos^2 f)(1 + e \cos f)^{-1}.$$

Now $-\mathbf{j} \cdot \mathbf{c} = \mathbf{i} \times \mathbf{k} \cdot \mathbf{c} = \mathbf{i} \cdot \mathbf{k} \times \mathbf{c} = \mathbf{i} \cdot \mathbf{n}$ so that

$$-\mathbf{j} \cdot \mathbf{c} = c \sin i \cos \Omega. \qquad (17.3)$$

Also, as the reader may demonstrate,

$$\boldsymbol{\alpha} \cdot \mathbf{j} = rc \left[-\sin(\omega + f)\sin \Omega + \cos(\omega + f)\cos \Omega \cos i \right].$$

$$(17.4)$$

Therefore, if we take the dot product of both sides of (16.8) with $\mathbf{j}$, the result is

$$\mathbf{j} \cdot \dot{\mathbf{c}} = -rF_\alpha \sin i \cos \Omega$$

$$+ rF_c \sin(\omega + f) \sin \Omega$$

$$- rF_c \cos(\omega + f) \cos \Omega \cos i,$$

which, according to (17.1) and (17.2), may also be written

$$-\mathbf{j} \cdot \dot{\mathbf{c}} = \dot{c} \sin i \cos \Omega - rF_c \sin(\omega + f) \sin \Omega$$

$$+ c \cos i \cos \Omega \frac{di}{dt} .$$

A direct differentiation of (17.3) shows agreement with the last equation, provided that

$$c\dot{\Omega} \sin i = rF_c \sin(\omega + f).$$

The computation of $\dot{\omega}$ starts with the observation that $\mathbf{n} \times \mathbf{e} = (\mathbf{k} \times \mathbf{c}) \times \mathbf{e} = (\mathbf{k} \cdot \mathbf{e})\mathbf{c}$, so that $ne \sin \omega = (\mathbf{k} \cdot \mathbf{e})c$, or

$$\mathbf{k} \cdot \mathbf{e} = e \sin i \sin \omega. \qquad (17.5)$$

In addition, we have the formula

$$\mathbf{k} \cdot \mathbf{r} = r \sin i \sin(\omega + f), \qquad (17.6)$$

easily obtained by substituting $c^{-2}(\boldsymbol{\alpha} \times \mathbf{c})$ for $\mathbf{r}$, and using the formula

$$\mathbf{k} \cdot \boldsymbol{\alpha} = cr \sin i \cos(\omega + f), \qquad (17.7)$$

derived at the beginning of the section. The remaining steps are these. Differentiate (17.5) and replace $\dot{\mathbf{e}}$, $\dot{e}$, di/dt by their equivalents obtained in this section and the preceding one. This yields an equation for $\dot{\omega}$ in terms of $(\mathbf{k} \cdot \dot{\mathbf{e}})$. Now dot multiply both sides of (16.9) with $\mathbf{k}$, substituting from (17.6) and (17.7). This gives us an evaluation of $(\mathbf{k} \cdot \dot{\mathbf{e}})$ which, on comparison with the preceding one, yields a formula for $\dot{\omega}$ given below.

There still remains the determination of $\dot{T}$. This we leave to the next section. In summary, we have found these formulas:

$$\dot{c} = rF_\alpha,$$

$$\mu c^{-1}\dot{e} = F_r \sin f + F_\alpha(e + 2\cos f + e\cos^2 f)(1 + e\cos f)^{-1},$$

$$\frac{di}{dt} = rc^{-1}F_c \cos(\omega + f), \qquad (17.8)$$

$$c\dot{\Omega} \sin i = rF_c \sin(\omega + f),$$

$$\dot{\omega} = -c\mu^{-1}e^{-1}(\cos f)F_r - rc^{-1}\cot i \sin(\omega + f)F_c$$

$$\qquad + (\mu ec)^{-1}(c^2 + r\mu)(\sin f)F_\alpha.$$

EXERCISE 17.1. Prove (17.4) by consulting Fig. 10. Recall that $\boldsymbol{\alpha} = \mathbf{c} \times \mathbf{r}$.

EXERCISE 17.2. Give a detailed proof of (17.6).

EXERCISE 17.3. Verify the formula for $\dot{\omega}$.

18. DISTURBED MOTION: GEOMETRIC EFFECTS

To complete the calculation summarized by (17.8) we now suppose that the undisturbed motion is elliptical. In that case, $0 < e < 1$ and $c^2 = \mu a(1 - e^2)$. Since $\dot{c}$ and $\dot{e}$ have already been found, it is easy to calculate $\dot{a}$ from this last equation. The result is

$$\dot{a} = 2a^2ec^{-1}(\sin f)F_r + 2a^2c\mu^{-1}r^{-1}F_\alpha. \qquad (18.1)$$

Since $n = \mu^{1/2}a^{-3/2}$, we know that $\dot{n} = -\frac{3}{2}na^{-1}\dot{a}$.

Finally, we determine $\dot{T}$. At the instant t, let a, n, e be the customary quantities associated with location on an elliptic orbit. Then we have the equations

$$r = \frac{c^2/\mu}{1 + e \cos f}$$

and

$$\dot{r} = \frac{\mu e}{c} \sin f.$$

Let us define u and T by the equations

$$r = a(1 - e \cos u),$$

$$n(t - T) = u - e \sin u.$$

Now $r\dot{r} = ce \sin f/(1 + e \cos f)$. But from the definition of u we have $\sin f/(1 + e \cos f) = \sin u/\sqrt{1 - e^2}$. Thus $r\dot{r} = \sqrt{\mu a}\ e \sin u$. Thus the equation (10.3) still holds in perturbed motion, although the relation $\dot{u} = \sqrt{\mu/a}\ r^{-1}$ is no longer valid. If we use this fact, then differentiation of the equation defining u yields

$$r^{-1}\sqrt{\mu a}\ e \sin u = \dot{a}(1 - e \cos u) + a(e\dot{u} \sin u - \dot{e} \cos u).$$

The equation defining T gives

$$\dot{n}(t - T) + n(1 - \dot{T}) = (1 - e \cos u)\dot{u} - \dot{e} \sin u.$$

If we (i) eliminate $\dot{u}$ between the last equations; (ii) replace $\dot{n}$ by $-\frac{2}{2} na^{-1}\dot{a}$, $1 - e \cos u$ by ra^{-1}, $\sin u$ by $r(1 - e^2)^{-1/2}a^{-1} \sin f$; (iii) solve for $\dot{T}$, the result is

$$\dot{T}\mu e \sin f = a^{-1}\left[rc - \tfrac{3}{2}\ \mu e(t - T)\sin f\right]\dot{a} - ac(\cos f)\dot{e}.$$

$$(18.2)$$

It is important to observe which of the elements is affected by which of the components F_r, F_c, F_α. The results are tabulated below.

$$F_r \quad \text{affects} \quad \dot{e}, \dot{\omega}, \dot{a}, \dot{T},$$

$$F_c \quad \text{affects} \quad \frac{di}{dt}, \dot{\Omega}, \dot{\omega},$$

$$F_\alpha \quad \text{affects} \quad \dot{c}, \dot{e}, \dot{\omega}, \dot{a}, \dot{T}.$$

We shall be content to illustrate their use with a simple example. Suppose a mass moving in an elliptic orbit, $0 < e < 1$, encounters a region of resistance, due, say, to atmosphere. The force will sometimes be of the form $\mathbf{F} - q\mathbf{v}$, where q is positive, although not necessarily a constant. What is the effect on the elements of the orbit? To solve the problem, observe that, according to (16.6),

$$\mathbf{F} = -qA\mathbf{r} - qB\boldsymbol{\alpha}.$$

Therefore, $F_r = -qAr$, $F_\alpha = -qB\alpha = -qBrc$. Using the computed values of A and B, we find that $F_r = -q\mu ec^{-1}\sin f$, $F_\alpha = -qr^{-1}c$. Clearly, $F_c = 0$. Substituting into (17.8) and (18.1), we get for the geometric elements of the orbit

$$\dot{e} = -2q(e + \cos f),$$

$$\frac{di}{dt} = 0,$$

$$\dot{\Omega}\sin i = 0,$$

$$\dot{\omega} = -2qe^{-1}\sin f,$$

$$\dot{a} = -2qa^2c^{-2}\mu(1 + 2e\cos f + e^2).$$

The following conclusions are immediate. The eccentricity e increases if $e + \cos f < 0$ and decreases if $e + \cos f > 0$.

(These correspond, respectively, to the left and right half of the ellipse.) The inclination is unchanged. The longitude of nodes Ω is unchanged, provided $i \neq 0$. (If $i = 0$, the angle Ω is, of course, undefined.) The amplitude of pericenter ω decreases in the upper half of the ellipse and increases in the lower half. The major axis always decreases.

EXERCISE 18.1. Verify that the formulas (17.8) and (18.1) are dimensionally correct. Use L for r and a, L^3T^{-2} for μ (why?), L^2T^{-1} for c, LT^{-2} for components of force, while e and angles are dimensionless.

EXERCISE 18.2. Find analogous formulas for the variation of the elements when the force $\mathbf{F}$ is decomposed in the directions $\mathbf{c}$, $\mathbf{v}$, $\mathbf{c} \times \mathbf{v}$.

CHAPTER 2

INTRODUCTION TO THE n-BODY PROBLEM

1. THE BASIC EQUATIONS: CONSERVATION OF LINEAR MOMENTUM

In the n-body problem (better, the n-particle problem) we are concerned with the motion of n mass particles of masses m_i, $i = 1, \ldots, n$ respectively, attracting one another in pairs with the force $Gm_j m_k r_{jk}^{-2}$ where r_{jk} is the distance between the kth and jth particle. We suppose that $n \geqslant 2$.

Let O represent an origin fixed in space and let $\mathbf{r}_i$, $\mathbf{v}_i$ denote the position and velocity vectors of the ith particle. Then, by Newton's second law, the kth particle satisfies the equation

$$m_k \ddot{\mathbf{r}}_k = \sum_{\substack{j=1 \\ j \neq k}}^{n} \frac{Gm_j m_k}{r_{jk}^2} \frac{\mathbf{r}_j - \mathbf{r}_k}{r_{jk}}, \qquad k = 1, \ldots, n, \quad (1.1)$$

where the right-hand side represents the total force exerted on the kth particle by the remaining $(n-1)$ particles.

We take for granted an important existence theorem governing the solutions of Eq. (1.1). The proof can be found in Sec. 409 of the book of Wintner referred to in the footnote, p. 35. Let the vectors $\mathbf{r}_i$, $\mathbf{v}_i$ be given at some instant $t = 0$ at which all the $\mathbf{r}_{jk}$ are positive. These we call the *initial data*. Let $r(t)$ denote the *smallest* of the distances r_{jk} at time t. Then there exists a unique set of n vector functions $\mathbf{r}_i(t)$ and a largest interval of time $-t_2 < t < t_1$ containing the instant $t = 0$ such that

(i) $\mathbf{r}_i(t)$ satisfies the differential Eq. (1.1) for $-t_2 < t < t_1$,

(ii) $\mathbf{r}_i(t)$ and $\mathbf{v}_i(t) = \dot{\mathbf{r}}_i(t)$ agree with the initial data when $t = 0$. Moreover,

(iii) if the interval $-t_2 < t < t_1$ is not the interval $-\infty < t < \infty$, then $r(t) \to 0$ as $t \to t_1$ if t_1 is finite and $r(t) \to 0$ as $t \to -t_2$ if t_2 is finite.

It must *not* be supposed in case (iii) that a collision occurs when $r \to 0$. It has never been proved (unless $n = 2$ or $n = 3$) that the only obstruction to the existence of the motion for all time is a collision of two or more particles. To put it another way, the fact that the minimum spacing $r(t)$ between particles becomes zero in no way implies that a particular pair collides.

The system (1.1) is of order $6n$ since there are n vector equations each of order 2, or $3n$ scalar equations each of order 2. One should anticipate $6n$ constants associated with the motion, that is, $6n$ functions of the $\mathbf{r}_i$, $\mathbf{v}_i$ and t which remain constant during the motion. These are known if $n = 2$ (see Sec.14, Chap.1), but in the general case only ten are known.

Six of the constants are easy to derive simply by adding the equations together. Clearly the double sum*

$$\sum_k \sum_{\substack{j \\ j \neq k}} \frac{Gm_jm_k}{r_{jk}^2} \frac{\mathbf{r}_j - \mathbf{r}_k}{r_{jk}}$$

---

* Hereafter we use Σ_k to mean $\displaystyle\sum_{k=1}^{n}$

vanishes, since for each occurrence of a term $\mathbf{r}_m - \mathbf{r}_n$ the term $\mathbf{r}_n - \mathbf{r}_m$ also occurs to cancel it. Therefore $\sum_k m_k \ddot{\mathbf{r}}_k = 0$. Now let M equal $\sum_k m_k$, the total mass, and let $\mathbf{r}_c$ denote the center of mass $M^{-1} \sum_k m_k \mathbf{r}_k$. Then $\ddot{\mathbf{r}}_c = 0$. Consequently $\mathbf{r}_c = \mathbf{a}t + \mathbf{b}$, where $\mathbf{a}$ and $\mathbf{b}$ are constant vectors, computable from the initial conditions. This last equation is the principle of conservation of linear momentum: *the center of mass moves uniformly in a straight line.* The vectors $\mathbf{a}$ and $\mathbf{b}$ provide six of the ten constants.

Since the motion of the center of mass is determined, the vital problem becomes the determination of the motion *relative* to the center of mass. For this purpose, it is convenient to move the origin to the center of mass by replacing each $\mathbf{r}_i$ by $\mathbf{r}_i - \mathbf{r}_c$. Because $\ddot{\mathbf{r}}_c = 0$, the Eqs. (1.1) are unaltered by the change. For this reason *we shall simply assume from now on that the center of mass of the system is fixed at the origin.* In other words, the system of Eqs. (1.1) carries with it the side condition

$$\sum_k m_k \mathbf{r}_k = 0, \qquad -t_2 < t < t_1, \qquad (1.2)$$

and hence also the condition

$$\sum_k m_k \mathbf{v}_k = 0, \qquad -t_2 < t < t_1. \qquad (1.3)$$

This provides six conditions to which the Eqs. (1.1) are subject, so that the system is of the order $6n - 6$.

EXERCISE 1.1.   Three equal masses start at rest from the vertices of an equilateral triangle. Prove they will collide, and find out when.

EXERCISE 1.2.   Explain why the condition $r \to 0$ does not imply a collision of two or more of the masses.

EXERCISE 1.3.   Suppose the law of attraction is $f(r) = \mu r$ rather than the inverse square law. Show that, as above, the origin can be moved to the center of mass, and that the resulting equations of motion become independent and can be solved completely.

2. THE CONSERVATION OF ENERGY: THE LAGRANGE-JACOBI FORMULA

We return to the Eqs. (1.1), assuming, henceforth, that the center of mass is fixed at the origin. Define the function U, the negative of the *potential energy*, by the equation

$$U = \sum_{1 \leqslant j < k \leqslant n} \frac{Gm_j m_k}{r_{jk}} . \qquad (2.1)$$

Since $r_{jk} = |\mathbf{r}_j - \mathbf{r}_k|$, the function U depends only on the positions $\mathbf{r}_1, \ldots, \mathbf{r}_n$ of the particles. In any Cartesian coordinate system fixed at O, the vector $\mathbf{r}_k$ will have components x_k, y_k, z_k, so that U can be regarded as a function of $x_1, y_1, z_1; x_2, y_2, z_2; \ldots; x_n, y_n, z_n$, a total of $3n$ real variables. By *the gradient of U in the direction* $\mathbf{r}_k$, we shall mean the vector having components

$$\left[\frac{\partial U}{\partial x_k}, \frac{\partial U}{\partial y_k}, \frac{\partial U}{\partial z_k} \right].$$

It is convenient to denote this vector by $\partial U / \partial \mathbf{r}_k$. In general, if $f(\mathbf{a}_1, \mathbf{a}_2, \ldots, \mathbf{a}_n)$ is a function of n vectors, we denote by $\partial f / \partial \mathbf{a}_k$ the vector

$$\frac{\partial f}{\partial \mathbf{a}_k} = \left[\frac{\partial f}{\partial \alpha_k}, \frac{\partial f}{\partial \beta_k}, \frac{\partial f}{\partial \gamma_k} \right],$$

where $\alpha_k, \beta_k, \gamma_k$ are the components of $\mathbf{a}_k$ in a Cartesian coordinate system. This seems more suggestive than the customary symbols $\Delta_k U$ or $\text{grad}_k U$.

It is now readily verified that the Eqs. (1.1) become

$$m_k \ddot{\mathbf{r}}_k = \frac{\partial U}{\partial \mathbf{r}_k} . \tag{2.2}$$

It follows from this that

$$\sum_k m_k \dot{\mathbf{r}}_k \cdot \ddot{\mathbf{r}}_k = \sum_k \frac{\partial U}{\partial \mathbf{r}_k} \cdot \frac{d\mathbf{r}_k}{dt} . \tag{2.3}$$

The right-hand side is clearly the total derivative of U with respect to t, since it can also be written

$$\sum_k \left[\frac{\partial U}{\partial x_k} \frac{dx_k}{dt} + \frac{\partial U}{\partial y_k} \frac{dy_k}{dt} + \frac{\partial U}{\partial z_k} \frac{dz_k}{dt} \right].$$

Therefore, because $v_k^2 = (\dot{\mathbf{r}}_k \cdot \dot{\mathbf{r}}_k)$, (2.3) can be written

$$\frac{d}{dt} \frac{1}{2} \sum_k m_k v_k^2 = \dot{U}.$$

Denote by T the *kinetic energy* $\frac{1}{2}\sum_k m_k v_k^2$, Then $\dot{T} = \dot{U}$ or

$$T = U + h, \tag{2.4}$$

when h is a constant, the total *energy*.

An extremely important form of this law is *the La-grange-Jacobi identity*. Define the moment of inertia $2I$ of the system by the formula

$$I = \tfrac{1}{2}\sum_k m_k r_k^2 = \tfrac{1}{2}\sum_k m_k(\mathbf{r}_k \cdot \mathbf{r}_k).$$

Differentiate the extreme members of this twice with respect to t. The result is

$$\ddot{I} = \sum_k m_k(\mathbf{v}_k \cdot \mathbf{v}_k) + \sum_k \mathbf{r}_k \cdot m_k \ddot{\mathbf{r}}_k, \tag{2.5}$$

or, by the Eqs. (1.1),

$$\ddot{I} = \sum_k m_k v_k^2 + \sum_k \sum_{j \neq k} \frac{Gm_j m_k}{r_{jk}^3} \left[(\mathbf{r}_j \cdot \mathbf{r}_k) - r_k^2 \right]$$

$$= 2T + \tfrac{1}{2} \sum_k \sum_{j \neq k} \frac{Gm_j m_k}{r_{jk}^3} \left[r_j^2 - r_k^2 - r_{jk}^2 \right].$$

Therefore

$$\ddot{I} - 2T = \tfrac{1}{2} \sum_k \sum_{j \neq k} \frac{Gm_j m_k}{r_{jk}^3} r_j^2 - \tfrac{1}{2} \sum_k \sum_{j \neq k} \frac{Gm_j m_k}{r_{jk}^3} r_k^2$$

$$- \tfrac{1}{2} \sum_k \sum_{j \neq k} \frac{Gm_j m_k}{r_{jk}} .$$

The first two terms on the right cancel one another, since they become identical if in the first one j and k are interchanged. The last term, by (2.1), is simply $-U$. Therefore $\ddot{I} = 2T - U$. By (2.4)

$$\ddot{I} = T + h = U + 2h. \tag{2.6}$$

EXERCISE 2.1. Write (2.5) as

$$\ddot{I} = 2T + \sum_k \mathbf{r}_k \cdot \frac{\partial U}{\partial \mathbf{r}_k}$$

using (2.2). Conclude that $\sum_k \mathbf{r}_k \cdot \partial U / \partial \mathbf{r}_k = -U$. Show from this that there is no arrangement of the n attracting particles so that they all remain at rest.

EXERCISE 2.2. Assuming that the particles move for all time $t > 0$ without obstruction, show by (2.6) that if $h > 0$ then $I \to \infty$ as $t \to \infty$. Conclude that at least one distance r_k cannot remain bounded. This does *not* say that some r_k becomes infinite.

*EXERCISE 2.3. Define a function T_1 of n vectors $\mathbf{p}_1, \mathbf{p}_2, \ldots, \mathbf{p}_n$ by $\frac{1}{2}\sum_k m_k^{-1} p_k^2$. Prove that

$$\frac{\partial T_1}{\partial \mathbf{p}_k} = m_k^{-1} \mathbf{p}_k.$$

*EXERCISE 2.4. Define the function $H(\mathbf{r}_1, \ldots, \mathbf{r}_n; \mathbf{p}_1, \ldots, \mathbf{p}_n)$ by $H = T_1 - U$. This is a function of $2n$ vectors or $6n$ scalars. Using the preceding exercise, show that the equations of motion (2.2) can be written in the Hamiltonian form

$$\dot{\mathbf{r}}_k = \frac{\partial H}{\partial \mathbf{p}_k}$$

$$\dot{\mathbf{p}}_k = -\frac{\partial H}{\partial \mathbf{r}_k}.$$

EXERCISE 2.5. Prove from the preceding exercise that $dH/dt = 0$ for a motion of the system. Derive (2.4) as a consequence.

*EXERCISE 2.6. Define r as in Sec. 1 of this chapter. Prove by (2.1) that $U \leqslant Ar^{-1}$, where A is a constant depending only on the masses. Conclude from the existence theorem of Sec. 1 that a solution of the equations of motion exists for $t > 0$, provided $U < \infty$ for $t > 0$.

3. THE CONSERVATION OF ANGULAR MOMENTUM

The constancy of the energy reduces the system from order $6n - 6$ to $6n - 7$. We now make a further reduction of three, to order $6n - 10$, by introducing the angular momentum $\mathbf{c}$. Cross multiply each side of (1.1) by $\mathbf{r}_k$ and sum on k. Since $\mathbf{r}_k \times \mathbf{r}_k = 0$, we conclude that

$$\sum_k m_k(\mathbf{r}_k \times \ddot{\mathbf{r}}_k) = \sum_k \sum_{j \neq k} \frac{Gm_j m_k}{r_{jk}^3} (\mathbf{r}_j \times \mathbf{r}_k).$$

The right-hand side vanishes because, with each occurrence of a term $\mathbf{r}_m \times \mathbf{r}_s$, the term $\mathbf{r}_s \times \mathbf{r}_m$ also occurs to cancel it. Therefore, the left-hand side is zero. Integration yields

$$\mathbf{c} = \sum_k m_k(\mathbf{r}_k \times \mathbf{v}_k), \tag{3.1}$$

where the constant $\mathbf{c}$ is the *angular momentum*.

Recall that if $a_1, \ldots, a_n; b_1, \ldots, b_n$ are $2n$ real numbers, then the quantities A, B, C, defined by

$$A = \sum_k a_k^2, \qquad B = \sum_k b_k^2, \qquad C = \sum_k a_k b_k,$$

are related by Cauchy's inequality* $C^2 \leqslant AB$. This has an important consequence, Sundman's inequality:

$$c^2 \leqslant 4I(\ddot{I} - h). \tag{3.2}$$

To prove (3.2) start with (3.1). This tells us that the length c of $\mathbf{c}$ satisfies the inequality

$$c \leqslant \sum_k m_k |\mathbf{r}_k \times \mathbf{v}_k|$$
$$\leqslant \sum_k m_k r_k v_k = \sum_k (\sqrt{m_k}\, r_k)(\sqrt{m_k}\, v_k).$$

Therefore, by Cauchy's inequality,

$$c^2 \leqslant \sum_k m_k r_k^2 \sum_k m_k v_k^2 = (2I)(2T).$$

---

* See Ex. 3.1.

According to (2.6), an immediate consequence is the inequality (3.2).

EXERCISE 3.1.   Prove Cauchy's inequality, starting with the obvious inequality

$$\sum_k (Ba_k - Cb_k)^2 \geq 0.$$

(The cases $B = 0$ and $B \neq 0$ must be treated separately.)

*EXERCISE 3.2.   (For use in the next section.) Let $f(x)$ be a twice-differentiable function defined on an interval $a \leq x \leq b$. Assume that $f > 0$, $f'' > 0$ on this interval and that $f(b) = 0$. Draw a graph to convince yourself that $f' \leq 0$ and prove it.

4. SUNDMAN'S THEOREM OF TOTAL COLLAPSE

In this section we shall study ths possibility that the system of particles suffers *total collapse*. By this we mean that *all* the particles come together at the same time, finite or infinite. We begin by writing the moment of inertia $2I$ in a new form. Since

$$\sum_j m_j(\mathbf{r}_j - \mathbf{r}_k)^2 = \sum_j m_j r_j^2 - 2\mathbf{r}_k \cdot \sum_j m_j \mathbf{r}_j + \sum_j m_j r_k^2,$$

we conclude from (1.2) that

$$\sum_j m_j(\mathbf{r}_j - \mathbf{r}_k)^2 = 2I - 0 + M r_k^2,$$

where M is the total mass. Multiply each side by m_k and sum. Since $r_{jk}^2 = (\mathbf{r}_j - \mathbf{r}_k)^2$, the result is

$$\sum_k \sum_j m_j m_k r_{jk}^2 = 2IM + M(2I) = 4IM.$$

On the left-hand side we can delete the term for which

$j = k$, since then $r_{jk} = 0$. Therefore

$$\sum_{1 \leqslant j < k \leqslant n} m_j m_k r_{jk}^2 = 2IM. \tag{4.1}$$

Since total collapse means that all r_{jk} become zero simultaneously, it follows from (4.1) that total collapse means that $I \to 0$, or that all particles simultaneously meet the origin.

First we show that *if total collapse is to occur, it will not take forever to happen*. In other words, $I \to 0$ as $t \to \infty$ is impossible. To prove this, return to (2.1). If all $r_{jk} \to 0$ as $t \to \infty$, then $U \to \infty$. Therefore, by (2.6), $\ddot{I} \to \infty$ because h is constant. This means that from some time on $\ddot{I} \geqslant 1$, say for $t \geqslant t_1$. Integrate both sides and we get $I \geqslant \frac{1}{2} t^2 + At + B$, where A and B are constants. Therefore, as $t \to \infty$, $I \to \infty$. This contradicts $I \to 0$.

Now we prove the more profound theorem* of Sundman. *Total collapse cannot occur unless the angular momentum is zero.* To prove this, suppose that $I \to 0$ as $t \to t_1$, where t_1 is finite. Just as before, $U \to \infty$ and $\ddot{I} \to \infty$ as $t \to t_1$. Therefore (we assume $t_1 > 0$ and let the reader modify the proof if $t_1 < 0$), $\ddot{I} > 0$ for some interval of time $t_2 \leqslant t \leqslant t_1$. Since $\ddot{I} > 0$, it follows from Ex. 3.2 that $-\dot{I} \geqslant 0$ for $t_2 \leqslant t \leqslant t_1$. Now multiply both sides of the inequality (3.2) by the positive number $-\dot{I} I^{-1}$. Therefore,

$$-\tfrac{1}{4} c^2 \dot{I} I^{-1} \leqslant h\dot{I} - \dot{I}\ddot{I}.$$

Integrate both sides with respect to t, for $t \geqslant t_2$. Then

$$\tfrac{1}{4} c^2 \log I^{-1} \leqslant hI - \tfrac{1}{2}\dot{I}^2 + K \leqslant hI + K,$$

---

* This was known to Weierstrass, who never published a proof.

where K is a constant of integration, so that

$$\tfrac{1}{4}c^2 \leqslant \frac{hI + K}{\log I^{-1}} .$$

Now let $t \to t_1$. Since $I \to 0$, it follows that $c^2 \to 0$. But c is a constant. Therefore $c = 0$.

5. THE VIRIAL THEOREM

We now assume (see Ex. 2.6) that the system moves from the instant $t = 0$ so that U remains finite. There is a classical result, called the Virial Theorem, which states that if I and T remain bounded for $t > 0$, then the two limits

$$\hat{T} = \lim_{t \to \infty} \frac{1}{t} \int_0^t T(\tau)d\tau, \qquad \hat{U} = \lim_{t \to \infty} \frac{1}{t} \int_0^t U(\tau)d\tau,$$

exist and $2\hat{T} = \hat{U}$. Since $T = U + h$, it follows that if one of the limits exists, so does the other and $\hat{T} = \hat{U} + h$. Therefore the conclusion $2\hat{T} = \hat{U}$ is equivalent to

$$\hat{T} = -h. \tag{5.1}$$

In this section we shall prove a sharper form of the theorem which does not require boundedness.*

THEOREM: *The statement* $\hat{T} = -h$ *is true if and only if*

$$\lim_{t \to \infty} t^{-2}I(t) = 0. \tag{5.2}$$

---

* H. Pollard, *A Sharp Form of the Virial Theorem, Bulletin of the American Mathematical Society,* LXX, (1964), 703–5.

We start with the Lagrange-Jacobi formula $\ddot{I} = T + h$. Integrate once and divide by t. Then

$$t^{-1}\dot{I} = t^{-1}\int_0^t T(\tau)d\tau + h + t^{-1}k, \qquad (5.3)$$

where k is a constant. Now let $t \to \infty$. From the definition of $\hat{T}$, the assertion (5.1) means that the right-hand side of (5.3) approaches zero, and hence the left-hand side also. Therefore (5.1) holds if and only if

$$\lim_{t \to \infty} t^{-1}\dot{I} = 0. \qquad (5.4)$$

It remains to show that each of (5.2) and (5.4) implies the other.

First suppose that (5.4) is true. Then, for each $\epsilon > 0$, it follows that $\dot{I} < \epsilon t$, provided t is large. Integrate both sides of the inequality. Then, $I < \epsilon(t^2/2) + At + B$, where A and B are constants. Therefore $t^{-2}I < (\epsilon/2) + t^{-1}A + t^{-2}B$. The last two terms can be made less than $\epsilon/2$ by taking t sufficiently large. Hence $t^{-2}I < \epsilon$ for large t. This proves (5.2).

Now let (5.2) be true. There is a theorem of Landau[*] which says that (5.4) is an immediate consequence, provided it is true that $\ddot{I} \geqslant -M$ for some finite number M. But $\ddot{I} = T + h$. Since $T \geqslant 0$, $\ddot{I} \geqslant h$, and the proof is finished.

EXERCISE 5.1.   Show that in the case of two bodies, the relation (5.1) holds if and only if $h \leqslant 0$. Show also that, in that case, if $h > 0$ then $\hat{T} = h$, $\hat{U} = 0$.

EXERCISE 5.2.   Prove that for a system of n bodies, the relation $\hat{T} = 0$ always implies $h = 0$. Suggestion:

_____

[*] For a proof see D. V. Widder, *The Laplace Transform*, Princeton University Press, 1942, p. 143.

since $\hat{T} = \hat{U} + h \geqslant 0$, it follows that $h \leqslant 0$. Now use (5.3) to conclude that $t^{-1}\dot{I} \to h$, $t^{-2}I \to \frac{1}{2}h$, so that $h \geqslant 0$.

6. GROWTH OF THE SYSTEM

We have seen in the case of the two-body problem that these cases occur: if $h < 0$, the system is bounded, that is, the distance r between the masses is bounded; if $h = 0$, the distance r grows like $|t|^{2/3}$ as $|t| \to \infty$ and if $h > 0$, r grows like $|t|$ as $|t| \to \infty$. The corresponding problems for three or more bodies is very difficult and we shall only obtain some elementary conclusions. It will be assumed that U remains finite.

First we reconsider the function

$$U = \sum \frac{Gm_j m_k}{r_{jk}},$$

where the sum is taken over the indices such that $1 \leqslant j < k \leqslant n$. Since $r \leqslant r_{jk}$, it follows that $U \leqslant A/r$, where A depends only on the masses.

Here is a simple consequence. Suppose $h < 0$. Then $T = U - |h|$. Since $T \geqslant 0$, we get $U \geqslant |h|$. Therefore $A/r \geqslant |h|$, or $r \leqslant A|h|^{-1}$. *If the energy is negative, the minimum distance is bounded.* The converse is false. In general, there is no simple relation between the growth of the system and the sign of the energy.

On the other hand, let m, m' be the two smallest masses. Then

$$U \geqslant \sum \frac{Gmm'}{r_{jk}} = Gmm' \sum \frac{1}{r_{jk}}.$$

Now, at any particular instant, r *is* one of the r_{jk}, so the sum on the right contains the term $1/r$. Therefore $U \geq Gmm'/r$. In summary,

$$B \leq rU \leq A, \tag{6.1}$$

where A and B are positive constants depending only on the masses. This says, roughly, that U^{-1} is a measure of r, the minimum spacing between particles.

We have shown that

$$I = \frac{1}{2M} \sum_{1 \leq j < k \leq n} m_j m_k r_{jk}^2.$$

Now denote by R the *maximum* of the r_{jk} at time t. Then $I \leq A_1 R^2$, where A_1 depends only on the masses. Arguing as in the preceding paragraph, let m, m' be the smallest masses. Then

$$I \geq \frac{mm'}{2M} \sum r_{jk}^2.$$

Since R is one of the r_{jk} at time t, $I \geq (mm'/2M)R^2$. Therefore

$$B_1 R^2 \leq I \leq A_1 R^2, \tag{6.2}$$

where A_1 and B_1 are positive constants determined by the masses. This means, roughly, that $\sqrt{I}$ is a measure of R, the maximum spacing between particles.

The question arises naturally of how rapidly a system can expand. We prove this elementary result: if $r \geq \delta > 0$, then $R \leq Mt$, where $\delta > 0$ and $M > 0$. This says that *if the particles do not get too close together at any time, then the maximum spacing cannot grow faster than the first power of t*. To prove it, we start once again with the formula $\ddot{I} = U + h$. Therefore $\ddot{I} \leq A/r + h$ or $\ddot{I} \leq A/\delta + h$. Integrating twice this says that $I \leq Dt^2$, where D is a constant. Therefore, by (6.2), $B_1 R^2 \leqslant Dt^2$, or $R \leqslant Mt$, where $M = (DB_1^{-1})^{1/2}$.

As a final application of these ideas, we repeat an argument used before. Since $\ddot{I} = U + 2h$, $\ddot{I} \geqslant 2h$. Suppose $h > 0$. Then $I \geqslant Et^2$, where E is a positive constant. Therefore $A_1 R^2 \geqslant Et^2$. Conclusion: *if $h > 0$, then R grows at least as fast as the first power of t.*

> EXERCISE 6.1. Use (6.2) to prove this form of the Virial Theorem: the statement $\hat{T} = -h$ is true if and only if $\lim\limits_{t \to \infty} t^{-1} R(t) = 0$.

> EXERCISE 6.2. Let ρ be the largest of the distances $r, \ldots, r_n$ of the masses from 0. Prove that
>
> $$m\rho^2 \leqslant I \leqslant M\rho^2,$$
>
> where m is the smallest mass. Conclude that R/ρ lies between two positive constants depending only on the masses. (Actually R/ρ never exceeds two. Why?) Show that the assertions $\hat{T} = -h$ and $\lim_{t \to \infty} t^{-1}\rho(t) = 0$ are equivalent.

7. THE THREE-BODY PROBLEM: JACOBI COORDINATES

In the special case $n = 3$, the equations of Sec. 1 become

$$m_1\ddot{\mathbf{r}}_1 = \frac{Gm_1m_2}{r_{12}^3}(\mathbf{r}_2 - \mathbf{r}_1) + \frac{Gm_3m_1}{r_{13}^3}(\mathbf{r}_3 - \mathbf{r}_1)$$

$$m_2\ddot{\mathbf{r}}_2 = \frac{Gm_1m_2}{r_{12}^3}(\mathbf{r}_1 - \mathbf{r}_2) + \frac{Gm_3m_2}{r_{23}^3}(\mathbf{r}_3 - \mathbf{r}_2) \quad (7.1)$$

$$m_3\ddot{\mathbf{r}}_3 = \frac{Gm_3m_1}{r_{13}^3}(\mathbf{r}_1 - \mathbf{r}_3) + \frac{Gm_3m_2}{\mathbf{r}_{23}^3}(\mathbf{r}_2 - \mathbf{r}_3).$$

Since $m_1\mathbf{r}_1 + m_2\mathbf{r}_2 + m_3\mathbf{r}_3 = 0$, one of the $\mathbf{r}_i$ can be eliminated. We prefer to proceed in another way. We shall consider the motion of m_2 relative to m_1 by use of the vector $\mathbf{r} = \mathbf{r}_2 - \mathbf{r}_1$ and of m_3 relative to the center of mass O' of m_1 and m_2. The location of this center is at $(m_1 + m_2)^{-1}(m_1\mathbf{r}_1 + m_2\mathbf{r}_2)$ or $-(m_1 + m_2)^{-1}m_3\mathbf{r}_3$. The position $\boldsymbol{\rho}$ of m_3 relative to this center is then $\mathbf{r}_3 + (m_1 + m_2)^{-1}m_3\mathbf{r}_3$ or $M\mu^{-1}\mathbf{r}_3$, where $\mu = m_1 + m_2$. Therefore $\boldsymbol{\rho} = M\mu^{-1}\mathbf{r}_3$.

It is easily verified, since $m_1\mathbf{r}_1 + m_2\mathbf{r}_2 + m_3\mathbf{r}_3 = 0$, that

$$\mathbf{r}_2 - \mathbf{r}_1 = \mathbf{r}; \quad \mathbf{r}_3 - \mathbf{r}_1 = \boldsymbol{\rho} + m_2\mu^{-1}\mathbf{r}; \quad \mathbf{r}_3 - \mathbf{r}_2 = \boldsymbol{\rho} - m_1\mu^{-1}\mathbf{r}.$$

We return to (7.1). Divide the first equation by m_1, the second by m_2 and subtract. The result is

$$\ddot{\mathbf{r}} = -\frac{G\mu}{r^3}\mathbf{r} + Gm_3\left[\frac{\boldsymbol{\rho} - m_1\mu^{-1}\mathbf{r}}{r_{23}^3} - \frac{\boldsymbol{\rho} + m_2\mu^{-1}\mathbf{r}}{r_{13}^3}\right]. \quad (7.2)$$

Now multiply the last equation of (7.1) by $M\mu^{-1}m_3^{-1}$. This time we get

$$\begin{aligned}
\ddot{\boldsymbol{\rho}} = &-\frac{MGm_1\mu^{-1}}{r_{13}^3}\left(\boldsymbol{\rho} + m_2\mu^{-1}\mathbf{r}\right) \\
&-\frac{MGm_2\mu^{-1}}{r_{23}^3}\left(\boldsymbol{\rho} - m_1\mu^{-1}\mathbf{r}\right).
\end{aligned} \quad (7.3)$$

The vectors $\mathbf{r}$ and $\boldsymbol{\rho}$ are called *Jacobi coordinates*.

We denote the relative velocity $\dot{\mathbf{r}}$ by $\mathbf{v}$ and $\dot{\boldsymbol{\rho}}$ by $\mathbf{V}$. Let $g_1 = m_1m_2\mu^{-1}$, $g_2 = m_3\mu M^{-1}$. It is readily verified that, in terms of the new coordinates $\mathbf{r}$ and $\boldsymbol{\rho}$,

$$\begin{aligned}
\mathbf{c} &= g_1(\mathbf{r} \times \mathbf{v}) + g_2(\boldsymbol{\rho} \times \mathbf{V}), \\
2I &= g_1r^2 + g_2\rho^2, \\
2T &= g_1v^2 + g_2V^2.
\end{aligned} \quad (7.4)$$

As a simple application, suppose that $c = 0$. Then $\mathbf{r} \cdot \boldsymbol{\rho} \times \mathbf{V} = 0$ and $\boldsymbol{\rho} \cdot \mathbf{r} \times \mathbf{v} = 0$. Therefore $\boldsymbol{\rho} \cdot \mathbf{r} \times \mathbf{V} = 0$ and $\mathbf{r} \cdot \mathbf{v} \times \boldsymbol{\rho} = 0$. Now let $\mathbf{u} = \mathbf{r} \times \boldsymbol{\rho}$. Then

$$\mathbf{u} \times \dot{\mathbf{u}} = (\mathbf{r} \times \boldsymbol{\rho}) \times (\mathbf{r} \times \mathbf{V}) + (\mathbf{r} \times \boldsymbol{\rho}) \times (\mathbf{v} \times \boldsymbol{\rho})$$

$$= (\mathbf{r} \cdot \mathbf{r} \times \mathbf{V})\boldsymbol{\rho} - (\boldsymbol{\rho} \cdot \mathbf{r} \times \mathbf{V})\mathbf{r}$$

$$+ (\mathbf{r} \cdot \mathbf{v} \times \boldsymbol{\rho})\boldsymbol{\rho} - (\boldsymbol{\rho} \cdot \mathbf{v} \times \boldsymbol{\rho})\mathbf{r} = 0.$$

Now according to the formula (2.2) of Chap. 1, it follows that $(d/dt)(\mathbf{u}/u) = 0$ when $u \neq 0$. Therefore, as long as $\mathbf{r} \times \boldsymbol{\rho} \neq 0$, the vector perpendicular to $\mathbf{r}$ and $\boldsymbol{\rho}$ is a constant. *It follows that all the motion is in one plane.* We leave it to the reader to draw the same conclusion if $\mathbf{r} \times \boldsymbol{\rho} = 0$ over an interval of time (Ex. 7.1).

EXERCISE 7.1. Complete the proof of Weierstrass' theorem: if $n = 3$, $c = 0$, all the motion takes place in a fixed plane. Conclude that if $n = 3$ a triple collision (total collapse) cannot occur unless all the motion takes place in a fixed plane. Suggestion: obtain a plane of motion by using $\mathbf{v}$ or $\mathbf{V}$ together with $\mathbf{r}$.

EXERCISE 7.2.   Verify formulas (7.4).

EXERCISE 7.3. Let H be a function of four independent vector variables $\mathbf{p}, \mathbf{P}, \mathbf{r}, \boldsymbol{\rho}$ defined by

$$H = \tfrac{1}{2}\frac{p^2}{g_1} + \tfrac{1}{2}\frac{P^2}{g_2} - \frac{Gm_1m_2}{r} - \frac{Gm_2m_3}{r_{23}} - \frac{Gm_3m_1}{r_{31}}.$$

Show that the Eqs. (7.2) and (7.3) can be written in the form

$$\frac{\partial H}{\partial \mathbf{p}} = \dot{\mathbf{r}}, \qquad \frac{\partial H}{\partial \mathbf{r}} = -\dot{\mathbf{p}},$$

$$\frac{\partial H}{\partial \mathbf{P}} = \dot{\boldsymbol{\rho}}, \qquad \frac{\partial H}{\partial \boldsymbol{\rho}} = -\dot{\mathbf{P}}.$$

Suggestion:

$$\frac{\partial}{\partial \mathbf{r}} r_{23}^{-1} = -\tfrac{1}{2} r_{23}^{-3} \frac{\partial r_{23}^2}{\partial \mathbf{r}}.$$

8. THE LAGRANGE SOLUTIONS

We seek a very special set of solutions of the three-body problem, namely those for which all three particles are moving uniformly in circles, in the same plane, and with the same angular velocity.

Introduce at O a fixed coordinate system x, y, z such that $z = 0$ is the plane of motion. Let $(x_k, y_k, 0)$ be the coordinates of the mass m_k. Then $\mathbf{r}_k = [x_k, y_k, 0]$ and the equations of motion (7.1) become

$$\ddot{x}_k = G \sum_{j \neq k} \frac{m_j}{r_{jk}^3} (x_j - x_k),$$

$$\ddot{y}_k = G \sum_{j \neq k} \frac{m_j}{r_{jk}^3} (y_j - y_k),$$

(8.1)

where $k = 1, 2, 3$, and each sum contains two terms.

Let the angular velocity of the particles in their plane of motion be ω. Introduce into that plane a coordinate system (ξ, η) which is rotating at angular velocity ω. In *this* coordinate system the particles are at rest. We transfer the Eqs. (8.1) to the new coordinate system, starting with the relations

$$x_k = \xi_k \cos \omega t - \eta_k \sin \omega t,$$

$$y_k = \xi_k \sin \omega t + \eta_k \cos \omega t.$$

(8.2)

We now differentiate each of these twice and substitute into (8.1). The following equations can then be derived for

ξ_k, η_k. When they have been solved, then (8.2) can be used to find (x_k, y_k).

$$\ddot{\xi}_k - 2\omega\dot{\eta}_k - \omega^2\xi_k = G \sum_{j \neq k} \frac{m_j}{r_{jk}^3} (\xi_j - \xi_k),$$

$$\ddot{\eta}_k + 2\omega\dot{\xi}_k - \omega^2\eta_k = G \sum_{j \neq k} \frac{m_j}{r_{jk}^3} (\eta_j - \eta_k), \tag{8.3}$$

where $k = 1, 2, 3$.

It is convenient to let $z_k = \xi_k + i\eta_k$, where $i = \sqrt{-1}$. Multiply the second of Eqs. (8.3) by i and add it to the first. We obtain

$$\ddot{z}_k + 2\omega i\dot{z}_k - \omega^2 z_k = G \sum_{j \neq k} \frac{m_j}{r_{jk}^3} (z_j - z_k), \tag{8.4}$$

where, of course, $r_{jk} = |z_j - z_k|$.

Since the particles are at rest in the rotating system, each $\dot{z}_k$ is identically zero. Therefore the positions z_1, z_2, z_3 we seek satisfy the equations

$$-z_k = \lambda \sum_{j \neq k} \frac{m_j}{r_{jk}^3} (z_j - z_k), \qquad k = 1, 2, 3,$$

where $\lambda = G\omega^{-2}$.

Let $\rho_1 = \lambda r_{23}^{-3}$, $\rho_2 = \lambda r_{31}^{-3}$, $\rho_3 = \lambda r_{12}^{-3}$. The first and third equations, written out in full, are

$$(1 - m_2\rho_3 - m_3\rho_2)z_1 + m_2\rho_3 z_2 + m_3\rho_2 z_3 = 0,$$

$$m_1\rho_2 z_1 + m_2\rho_1 z_2 + (1 - m_1\rho_2 - m_2\rho_1)z_3 = 0. \tag{8.5}$$

Since the center of mass is fixed at O, the missing equation can be replaced by

$$m_1 z_1 + m_2 z_2 + m_3 z_3 = 0.$$

There are two possibilities: (i) the points z_1, z_2, z_3 at some time t are not in a straight line; (ii) they are. In case (i) the coefficients of corresponding z_k in the preceding three equations are proportional. It follows immediately that $\rho_1 = \rho_2 = \rho_3 = 1/M$, where M is the total mass $m_1 + m_2 + m_3$. In other words, the only possible solution of the form (i) puts the masses at the vertices of an equilateral triangle of side $(GM\omega^{-2})^{1/3}$. It is important to observe that this is independent of the size of the masses, so that the center of mass and the center of the triangle need not coincide. This solution is due to Lagrange. Case (ii) will be treated in the next section.

EXERCISE 8.1. Prove in case (i) that the force on each mass passes through the origin.

EXERCISE 8.2. In case (i) compute the quantities T, U, I, h. Answer: $2T = U = -2h = 2\omega^2 I$, where $U = q(m_1 m_2 + m_2 m_3 + m_3 m_1)$ and $q = (G\omega)^{2/3} m^{-1/3}$.

9. EULER'S SOLUTION

Suppose now that z_1, z_2, z_3 at some instant t lie on a line L. Since L must contain the center of mass, it passes through O and we may as well suppose it is the ξ-axis so that all η_k vanish. By renumbering the masses, we can arrange that $\xi_1 < \xi_2 < \xi_3$ so that $r_{12} = \xi_2 - \xi_1$, $r_{23} = \xi_3 - \xi_2$, $r_{13} = \xi_3 - \xi_1$. The Eqs. (8.5) can be written

$$-\xi_1 = \lambda \left[\frac{m_2}{(\xi_2 - \xi_1)^2} + \frac{m_3}{(\xi_3 - \xi_1)^2} \right],$$

$$\xi_3 = \lambda \left[\frac{m_1}{(\xi_3 - \xi_1)^2} + \frac{m_2}{(\xi_3 - \xi_2)^2} \right],$$

(9.1)

where

$$m_1\xi_1 + m_2\xi_2 + m_3\xi_3 = 0. \qquad (9.2)$$

Now let $\xi_2 - \xi_1 = a$, $\xi_3 - \xi_2 = a\rho$, $\xi_3 - \xi_1 = a(1 + \rho)$. Equation (9.2) can be written in either of the forms

$$m_2 a + m_3 a(1 + \rho) = -M\xi_1,$$
$$m_1 a(1 + \rho) + m_2 a\rho = M\xi_3. \qquad (9.3)$$

Obtain $-\xi_1/\xi_3$ from each pair (9.1) and (9.3) by division. Equate the results to obtain

$$\frac{m_2 + m_3(1 + \rho)}{m_1(1 + \rho) + m_2\rho} = \frac{m_2 + m_3(1 + \rho)^{-2}}{m_1(1 + \rho)^{-2} + m_2\rho^{-2}}. \qquad (9.4)$$

The order of events is this. Suppose that ρ can be determined from this equation. Replace ξ_1 on the left-hand side of (9.1) from its value given by (9.3). We find that

$$a^3\left[m_2 + m_3(1 + \rho) \right] = \lambda M\left[m_2 + m_3(1 + \rho)^{-2} \right].$$

This determines a. Then (9.1) determines ξ_1 and ξ_3. Finally, $\xi_2 = a + \xi_1$.

This reduces the problem to the determination of positive values of ρ which satisfy (9.4). It can be written

$$(m_2 + m_3) + (2m_2 + 3m_3)\rho + (3m_3 + m_2)\rho^2$$
$$- (3m_1 + m_2)\rho^3 - (3m_1 + 2m_2)\rho^4 - (m_1 + m_2)\rho^5 = 0.$$

If $\rho = 0$, the left-hand side is positive; as $\rho \to \infty$ it approaches $-\infty$. Therefore it has a positive root. By Descartes' rule of signs it has at most one positive root. Hence there is a unique positive value of ρ which solves the problem. It is clear that, by renumbering the masses, two other solutions to the main problem can be obtained. These collinear solutions are due to Euler.

EXERCISE 9.1. Solve the problem explicitly if $m_1 = m_2 = m_3$.

10. THE RESTRICTED THREE-BODY PROBLEM

The three-body problem described by Eqs. (7.2) and (7.3) is a system of order twelve. Its equivalent formulation, given in Ex. 7.3, gives four vector equations (equal to twelve scalar equations), each of the first order. By use of Eq. (7.4) for the conservation of angular momentum and the conservation of energy, the system can be reduced by four, leaving eight. It is possible to eliminate the time from the eight, leaving a system of order seven, and finally, by a device due to Jacobi, it can be cut down to order six. Moreover, if the motion is planar, we use only two of the three dimensions of space and the order is reduced to four. This is the best that is known. After all these reductions, the problem is still extremely complicated and has kept mathematicians busy for over two hundred years.

We shall make an assumption which leads to a more tractable problem. It will be supposed that the mass m_3 is so small that it does not influence the motion of m_1 and m_2 (known as the *primaries*), but is affected by them in the usual way. Clearly, this is a sensible approximation to reality only if the path of m_3 does not come too close to m_1 or m_2. Mathematically what we do is to set $m_3 = 0$, or what is equivalent, $M = \mu$. The center of mass of the system is now the center of mass of the primaries. If we let

$r_{13} = \rho_1$ and $r_{23} = \rho_2$, the Eqs. (7.2) and (7.3) become, respectively,

$$\ddot{\mathbf{r}} = G\mu r^{-3}\mathbf{r} \qquad (10.1)$$

and

$$\ddot{\boldsymbol{\rho}} = -Gm_1\rho_1^{-3}(\boldsymbol{\rho} + m_2\mu^{-1}\mathbf{r})$$
$$-Gm_2\rho_2^{-3}(\boldsymbol{\rho} - m_1\mu^{-1}\mathbf{r}). \qquad (10.2)$$

The first equation can be solved completely by the methods of Chap. 1, and so $\mathbf{r}$ can be taken as a *known* solution of the two-body problem. Then the motion of m_3 is completely described by the single Eq. (10.2). This is called the *restricted* three-body problem. Because of our physical assumption on m_3, the customary conservation laws do not hold and we cannot use them to reduce the order of (10.2), which is six. We shall make the further assumption that all the motion occurs in one plane (the *planar* restricted problem), which makes the order four. Finally, we shall suppose that the primaries rotate uniformly around their center of mass (the *circular* planar restricted problem).

The mean motion n for the primaries, according to Eq. (10.1), is given by $\sqrt{G\mu}\, r^{-3/2}$, where r is the distance between the primaries. We may, therefore, use the rotating coordinate system described in Sec. 8 and illustrated in Fig. 11, with $\omega = n$. The primaries are at rest in this coordinate system and we shall place them on the ξ-axis. Equation (8.4) is applicable with $k = 3$. If we write z for z_3, ρ_1 for r_{13}, ρ_2 for r_{23}, it becomes

$$\ddot{z} + 2\omega i\dot{z} - \omega^2 z = Gm_1\rho_1^{-3}(z_1 - z)$$
$$+ Gm_2\rho_2^{-3}(z_2 - z). \qquad (10.3)$$

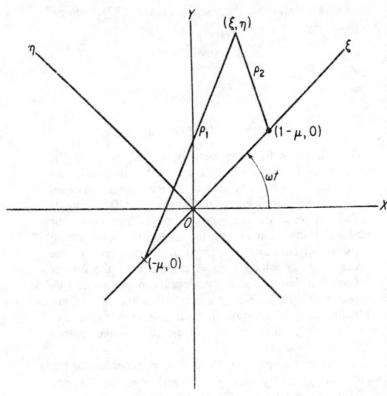

Figure 11

Remember that $\eta_1 = \eta_2 = 0$, so that $z_1 = \xi_1$, $z_2 = \xi_2$. Moreover, $z = \xi + i\eta$.

It is convenient to choose the unit of mass so that $m_1 + m_2 = 1$, of length so that $r = 1$ and of time so that $G = 1$. The lighter mass will be denoted by* μ and placed

---

* This use of μ should be distinguished from that occurring in the definition of *n*.

at ξ_2, to the right of the origin. Clearly, $\mu \leqslant 1/2$. Since $m_1\xi_1 + m_2\xi_2 = 0$ and $\xi_2 - \xi_1 = 1$, it follows that $\xi_1 = -\mu$, $\xi_2 = 1 - \mu$. The mass m_1 is located at $(-\mu, 0)$ and m_2 at $(1 - \mu, 0)$. Finally, observe that with all these choices of units, $n = \omega = 1$. The equation of motion in the rotating coordinate system has become

$$\ddot{z} + 2i\dot{z} - z = -\frac{(1 - \mu)(z + \mu)}{\rho_1^3} - \frac{\mu(z - 1 + \mu)}{\rho_2^3} . \quad (10.4)$$

The rest of this chapter will be devoted to a study of Eq. (10.4). As in Sec. 1, we ask the reader to accept an existence theorem; the same reference is applicable. Let initial values of z and $\dot{z}$ be given. Then there exists a unique function $z(t)$ and a largest interval of time $-t_2 < t < t_1$ containing the instant $t = 0$, such that Eq. (10.4) is satisfied and the initial conditions are met. Moreover, if either $-t_2$ or t_1 is finite, then either $\lim \rho_1 = 0$ or $\lim \rho_2 = 0$; that is, collision with one of the primary masses occurs.

EXERCISE 10.1. Derive Eq. (10.3) directly from (10.2). Suggestion: since the motion is planar, we can treat ρ and $\mathbf{r}$ as complex numbers. Let $\rho = ze^{i\omega t}$, $\mathbf{r} = e^{i\omega t}$.

EXERCISE 10.2. Show that (10.3), with $\omega = 0$, solves Ex. 2.1 of Chap. 1 in the case of Newtonian attraction.

*EXERCISE 10.3. Let

$$U = \frac{1 - \mu}{\rho_t} + \frac{\mu}{\rho_2} ,$$

where $\rho_1 = |z - z_1|$. Show that Eq. (10.4) can be written

$$\ddot{\xi} - 2\dot{\eta} - \xi = \frac{\partial U}{\partial \xi} ,$$

$$\ddot{\eta} + 2\dot{\xi} - \eta = \frac{\partial U}{\partial \eta} .$$
(10.5)

11. THE CIRCULAR RESTRICTED PROBLEM; THE JACOBI CONSTANT

It must not be supposed that the problem described in the last chapter is an artificial one. Two examples will serve to make a rather convincing argument that the problem is worth investigating.

Apart from the sun itself, the heaviest of all the planets is Jupiter, which moves in an ellipse of small eccentricity; call it a circle for a first approximation. There is a group of tiny planets, the Trojan asteroids, whose motion is controlled principally by the sun and Jupiter; a first approximation to their motion is given by a solution of the restricted problem with the sun and Jupiter as primaries.

As another example, consider the motion of the earth around the sun to be circular. Then these two play the role of the primaries and the moon is m_3, the small mass.

We turn to the main problem of investigating Eqs. (10.5). They are

$$\ddot{\xi} - 2\dot{\eta} - \xi = \frac{\partial U}{\partial \xi} ,$$

$$\ddot{\eta} + 2\dot{\xi} - \eta = \frac{\partial U}{\partial \eta} ,$$

where $U(\xi, \eta) = (1 - \mu/\rho_1) + (\mu/\rho_2)$.

If we define a new "potential" Φ by

$$\Phi(\xi, \eta) = \tfrac{1}{2}(\xi^2 + \eta^2) + U + \tfrac{1}{2}\mu(1 - \mu), \qquad (11.1)$$

the equations read, more simply,

$$\ddot{\xi} - 2\dot{\eta} = \frac{\partial \Phi}{\partial \xi}$$

$$\ddot{\eta} + 2\dot{\xi} = \frac{\partial \Phi}{\partial \eta}. \qquad (11.2)$$

The constant $\tfrac{1}{2}\mu(1 - \mu)$ appearing in the definition of Φ is of no importance in these equations, but is convenient later.

We have already explained in Sec. 10 that the usual conservation laws do not hold. But a substitute exists. Define the *Jacobi integral* as the expression $2\Phi - \dot{\xi}^2 - \dot{\eta}^2$. Multiply the first of Eqs (11.2) by $\dot{\xi}$, the second by $\dot{\eta}$ and add. The result is $\dot{\xi}\ddot{\xi} + \dot{\eta}\ddot{\eta} = d\Phi/dt$. Therefore

$$\dot{\xi}^2 + \dot{\eta}^2 = 2\Phi - C, \qquad (11.3)$$

where C is a constant, the so-called *Jacobi constant*. Equation (11.3) says that *the Jacobi integral remains equal to C during the motion*. It is clearly determined by the initial values ξ_0, η_0, $\dot{\xi}_0$, $\dot{\eta}_0$.

The system (11.2) can be written

$$\dot{\xi} = \alpha, \qquad\qquad \dot{\eta} = \beta,$$

$$\dot{\alpha} = 2\beta + \Phi_\xi, \qquad \dot{\beta} = -2\alpha + \Phi_\eta, \qquad (11.4)$$

which is of order four. Now divide the first two by the third to eliminate time. We find that

$$\frac{d\xi}{d\alpha} = \frac{\alpha}{2\beta + \Phi_\xi}, \qquad \frac{d\eta}{d\alpha} = \frac{\beta}{2\beta + \Phi_\xi}.$$

From (11.3) we know that $\alpha^2 + \beta^2 = 2\Phi - C$. This can be solved for β and the result substituted into the preceding pair to obtain equations of the form

$$\frac{d\xi}{d\alpha} = F(\xi, \eta, \alpha)$$

$$\frac{d\eta}{d\alpha} = G(\xi, \eta, \alpha),$$

a second order system. If the solution is given by $\xi = f(\alpha)$, $\eta = g(\alpha)$, then we proceed as follows. Since $\alpha = \dot{\xi} = f'(\alpha)\dot{\alpha}$, we can, in theory, determine $\alpha(t)$. Then $\xi = \xi_0 + \int_0^t \alpha(\tau)d\tau$, so that $\xi(t)$ is determined. Also $\dot{\eta} = \beta = g'(\alpha)\alpha' = \alpha g'(\alpha)/f'(\alpha)$. Therefore,

$$\eta = \eta_0 + \int_0^t \frac{\alpha g'(\alpha)}{f'(\alpha)}\, d\tau,$$

where $\alpha(\tau)$ must be substituted for α under the integral sign. In practice, this method is of no use since $f(\alpha)$ and $g(\alpha)$ are impossible to determine explicitly. Instead of pursuing this line of thought further, we shall seek some simple explicit solutions, analogous to those found in Secs. 8 and 9 for the unrestricted problem.

EXERCISE 11.1. In the theoretical solution described above the equation for $\dot{\beta}$ was never used. Why?

*EXERCISE 11.2. A more useful system than (11.4) can be obtained as follows. Write (11.2) as

$$\frac{d}{dt}(\dot{\xi} - \eta) = \dot{\eta} + \frac{\partial\Phi}{\partial\xi}$$

$$\frac{d}{dt}(\dot{\eta} + \xi) = -\dot{\xi} + \frac{\partial\Phi}{\partial\eta}.$$

This suggests the substitution $p = \dot{\xi} - \eta$, $P = \dot{\eta} + \xi$, so that

$$\frac{dp}{dt} = P - \xi + \frac{\partial \Phi}{\partial \xi}$$

$$\frac{dP}{dt} = -p - \eta + \frac{\partial \Phi}{\partial \eta}$$

$$\frac{d\xi}{dt} = p + \eta$$

$$\frac{d\eta}{dt} = P - \xi.$$

Now define $H(\xi, \eta; p, P) = \frac{1}{2}(p + \eta)^2 + \frac{1}{2}(P - \xi)^2 - \Phi(\xi, \eta)$ and verify that the system can be written in the form

$$\dot{\xi} = \frac{\partial H}{\partial p}, \qquad \dot{p} = -\frac{\partial H}{\partial \xi},$$

$$\dot{\eta} = \frac{\partial H}{\partial P}, \qquad \dot{P} = -\frac{\partial H}{\partial \eta}.$$

The initial values are ξ_0, η_0, $p_0 = \dot{\xi}_0 - \eta_0$, $\dot{P}_0 = \dot{\eta}_0 + \xi_0$.

12. EQUILIBRIUM SOLUTIONS

We seek solutions of the restricted problem for which the small mass m_3 remains at rest in the relative coordinate system. These are called *equilibrium* solutions. Since ξ and η are constant, the Eqs. (11.2) become simply

$$\frac{\partial \Phi}{\partial \xi} = \frac{\partial \Phi}{\partial \eta} = 0. \tag{12.1}$$

It is convenient to express Φ in terms of the so-called *bipolar* coordinates ρ_1 and ρ_2 of the point (ξ, η). Since $\rho_1^2 = (\xi + \mu)^2 + \eta^2$, $\rho_2^2 = (\xi - 1 + \mu)^2 + \eta^2$, we find that $\xi^2 + \eta^2 = (1 - \mu)\rho_1^2 + \mu\rho_2^2 - \mu(1 - \mu)$. From (11.1) and the definition of U, we get

$$\Phi = (1 - \mu)\left(\tfrac{1}{2}\rho_1^2 + \rho_1^{-1}\right) + \mu\left(\tfrac{1}{2}\rho_2^2 + \rho_2^{-1}\right). \quad (12.2)$$

The relations (12.1) become

$$(1 - \mu)\left[\rho_1 - \frac{1}{\rho_1^2}\right]\frac{\xi + \mu}{\rho_1} + \mu\left[\rho_2 - \frac{1}{\rho_2^2}\right]\frac{\xi - 1 + \mu}{\rho_2} = 0$$

$$(12.3)$$

$$(1 - \mu)\left[\rho_1 - \frac{1}{\rho_1^2}\right]\frac{\eta}{\rho_1} + \mu\left[\rho_2 - \frac{1}{\rho_2^2}\right]\frac{\eta}{\rho_2} = 0.$$

First suppose that $\eta \neq 0$. Then

$$(1 - \mu)\left[\rho_1 - \frac{1}{\rho_1^2}\right]\frac{1}{\rho_1} + \mu\left[\rho_2 - \frac{1}{\rho_2^2}\right]\frac{1}{\rho_2} = 0.$$

This means that the terms containing ξ in the first of Eqs. (12.3) drop out. In addition, a factor of $\mu(1 - \mu)$ cancels and we are left with

$$\left[\rho_1 - \frac{1}{\rho_1^2}\right]\frac{1}{\rho_1} - \left[\rho_2 - \frac{1}{\rho_2^2}\right]\frac{1}{\rho_2} = 0.$$

The only simultaneous solution of this equation and the preceding one is $\rho_1 = \rho_2 = 1$. Therefore, if $\eta \neq 0$, there are precisely two equilibrium solutions: the vertices of two equilateral triangles based on the line joining $(-\mu, 0)$ and

$(1 - \mu, 0)$. These are the points L_4 and L_5 indicated in Fig. 12.

On the other hand, if $\eta = 0$ the Eqs. (12.3) reduce to the single one

$$(1 - \mu)\left[\rho_1 - \frac{1}{\rho_1^2}\right]\frac{\xi + \mu}{\rho_1} + \mu\left[\rho_2 - \frac{1}{\rho_2^2}\right]\frac{\xi - 1 + \mu}{\rho_2} = 0,$$

$$(12.4)$$

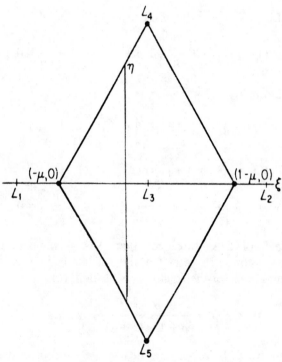

Figure 12

where $\rho_1 = |\xi + \mu|$, $\rho_2 = |\xi - 1 + \mu|$. There are three cases: $\xi < -\mu$, $-\mu < \xi < 1 - \mu$, $\xi > 1 - \mu$, in which we have, respectively,

(a) $\rho_1 = -\xi - \mu$, $\rho_2 = 1 - \xi - \mu$, $\rho_2 = 1 + \rho_1$;

(b) $\rho_1 = \xi + \mu$, $\rho_2 = 1 - \xi - \mu$, $\rho_2 = 1 - \rho_1$;

(c) $\rho_1 = \xi + \mu$, $\rho_2 = \xi + \mu - 1$, $\rho_2 = \rho_1 - 1$.

We can rewrite (12.4) in each of the cases as follows.

(a) Let $\rho_1 = \rho$, $\rho_2 = 1 + \rho$. Then

$$(1 - \mu)\left[\rho - \frac{1}{\rho^2}\right] + \mu\left[\rho + 1 - \frac{1}{(\rho + 1)^2}\right] = 0.$$

(b) Let $\rho_1 = \rho$, $\rho_2 = 1 - \rho$. Then

$$(1 - \mu)\left[\rho - \frac{1}{\rho^2}\right] = \mu\left[1 - \rho - \frac{1}{(1 - \rho)^2}\right].$$

(c) Let $\rho_2 = \rho$, $\rho_1 = 1 + \rho$. Then

$$(1 - \mu)\left[1 + \rho - \frac{1}{(1 + \rho)^2}\right] + \mu\left[\rho - \frac{1}{\rho^2}\right] = 0.$$

Each of these three equations has a single positive solution for ρ. In cases (a) and (c) this can be seen as follows. Each of the equations is of the form

$$F(\rho) = \frac{\rho - \rho^{-2}}{\rho + 1 - (\rho + 1)^{-2}} = -c,$$

where $c > 0$. It is easily verified that $F'(\rho) > 0$, so that F is strictly increasing. Moreover, $F(0+) = -\infty$, $F(1) = 0$.

Therefore F assumes the value $-c$ at precisely one value of ρ between 0 and 1. The solutions are denoted by L_1 in case (a) and by L_2 in case (c), as indicated in Fig. 12.

The case (b) is similar. Now the equation is

$$F_1(\rho) = \frac{1 - \rho - (1 - \rho)^{-2}}{\rho - \rho^{-2}} = \frac{1 - \mu}{\mu} \geqslant 1,$$

because $\mu \leqslant \frac{1}{2}$. The function $F_1(\rho)$ is increasing in the interval $\frac{1}{2} \leqslant \rho < 1$. Moreover, $F_1(\frac{1}{2}) = 1$, $F_1(1 -) = \infty$, so that F_1 assumes the value $1 - \mu/\mu$ precisely once in the interval $\frac{1}{2} \leqslant \rho < 1$. This means that the equilibrium point lies closer to the lighter mass than to the other, unless $\mu = \frac{1}{2}$; it is called L_3 (see Fig. 12).

The five points L_i are called *libration points*. The first three are called the *Euler points* and the last two the *Lagrange points*.

EXERCISE 12.1. Calculate the position of the five libration points in the case $\mu = \frac{1}{2}$, when the primaries have equal masses.

EXERCISE 12.2. On the assumption that the earth and the moon fulfill approximately the requirements of the primaries in the restricted three-body problem, what significance can be attached to the five L_i? Where are they located in this case? Assume $\mu = .012$.

EXERCISE 12.3. Show that the only solutions of the equation $\Phi = \frac{3}{2}$ are the libration points L_4 and L_5.

EXERCISE 12.4. Show that both $\partial\Phi/\partial\rho_1$ and $\partial\Phi/\partial\rho_2$ vanish at L_4, L_5, but neither does at L_1, L_2, L_3.

*EXERCISE 12.5.   Show that if the origin of coordinates is translated to L_4, the differential equations become

$$\ddot{x} - 2\dot{y} = x + \tfrac{1}{2}\rho^* + \frac{\partial U}{\partial x}$$

$$\ddot{y} + 2\dot{x} = y + \tfrac{1}{2}\sqrt{3} + \frac{\partial U}{\partial y},$$

where $\rho^* = 1 - 2\mu$ and

$$U = (1 - \mu)\left(1 + x + x^2 + \sqrt{3}\, y + y^2\right)^{-1}$$

$$+ \mu\left(1 - x + x^2 + \sqrt{3}\, y + y^2\right)^{-1}.$$

13.   THE CURVES OF ZERO VELOCITY

The equilibrium solutions are the only solutions of (11.2) that are known explicitly. However, by use of the Jacobi integral it is possible to derive some important general properties of all solutions. According to the formula (11.3), it is true that

$$v^2 = 2\Phi - C, \tag{13.1}$$

where v is the relative velocity $(\dot{\xi}^2 + \dot{\eta}^2)^{1/2}$, C is the Jacobi constant of the motion, and, in bipolar form,

$$2\Phi = (1 - \mu)(\rho_1^2 + 2\rho_1^{-1}) + \mu(\rho_2^2 + 2\rho_2^{-1}). \tag{13.2}$$

We shall consider the level curves $2\Phi = C$, which, in accordance with (13.1), are called the *curves of zero velocity*. It will now be proved that the minimum value of 2Φ is 3, so that no level curve exists when $C < 3$. We start with the assertion that if $0 \leqslant \mu \leqslant 1$, $A \geqslant 0$, $B \geqslant 0$, then

$$A\mu + B(1 - \mu) \geqslant \min(A, B).$$

For if $A \geqslant B$, then $A\mu + B(1 - \mu) \geqslant B\mu + B(1 - \mu) = B$ = min(A, B), and similarly if $A \leqslant B$. Therefore, by (13.2), $2\Phi \geqslant \min(\rho_1^2 + 2\rho_1^{-1}, \rho_2^2 + 2\rho_2^{-1})$. But the minimum of the function $x^2 + 2x^{-1}$ is 3, achieved when $x = 1$. Hence, $2\Phi \geqslant 3$. Clearly, this minimum is achieved only when $\rho_1 = \rho_2 = 1$, that is, at the Lagrange libration points. This, incidentally, solves Ex. 12.3.

We shall begin with $C = 3$, when the level curve $2\Phi = C$ consists only of the points L_4, L_5, and describe the shape of the curves as C increased. It will be supposed that $0 < \mu < \frac{1}{2}$. It is clear from the definition of Φ that the curves are symmetric in the axis $\eta = 0$, so we have drawn only the upper half of each. In the accompanying Fig. 13,

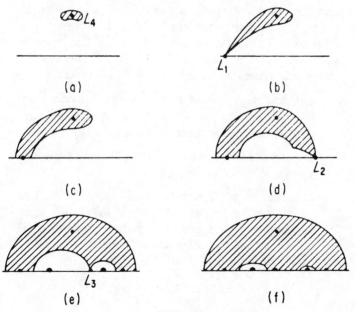

Figure 13

the shaded region corresponds to $2\Phi < C$. The drawings are schematic and do not pretend to any accuracy.

When C exceeds 3 slightly, the locus appears as a pair of curves surrounding L_4 and L_5 as in (a) of Fig. 13. As C increases, the left-hand edges of the curves join together at L_1, as in (b). After a transitional stage, as seen in (c), the curves join at L_2. This is shown in (d). At the next stage (e) there is a joining at L_3 and the primaries are surrounded. At the final stage the joining at L_3 disappears, and from this point on the general appearance is displayed by (f) in which the primary masses are enclosed by the inner curves.

The importance of the curves is this. Each locus $2\Phi = C$ divides the plane into the shaded region where $2\Phi < C$ and the unshaded region, where $2\Phi > C$. In view of Eq. (13.1), motion is impossible if $2\Phi < C$, since then $v^2 < 0$. Therefore, the shaded regions indicate for each value of the Jacobian constant C the positions in the ξ-η coordinate system where motion *cannot* take place.

> EXERCISE 13.1.   Show that if $\mu = \frac{1}{2}$, the curves are symmetric in the η-axis also. Stage (c) does not exist, and L_1 and L_3 are reached at the same time.

> EXERCISE 13.2.   Use the formula (13.2) to explain the general shape of the curves.

> EXERCISE 13.3.   Use the conclusions of Ex. 12.2 to determine the largest value of C below which an earth-moon trip is possible. Hint: configuration (e) makes such a trip impossible. Therefore C must be such that $2\Phi = C$ is satisfied by L_3.

CHAPTER 3

INTRODUCTION TO HAMILTON-JACOBI THEORY

1. CANONICAL TRANSFORMATIONS

We begin by recalling some basic facts from advanced calculus. Let the functions

$$y_k = y_k(x_1, \ldots, x_m), \qquad k = 1, \ldots, m \qquad (1.1)$$

denote a transformation of variables in an m-dimensional region. It will be supposed that each of the partial derivatives $\partial y_k / \partial x_l$ exists and is continuous. The matrix $\mathfrak{M}$ with entries $\partial y_k / \partial x_l$ (k = row index, l = column index) is known as the *Jacobian matrix* of the transformation; in more detail it is

$$\mathfrak{M} = \begin{bmatrix} \dfrac{\partial y_1}{\partial x_1} & \dfrac{\partial y_1}{\partial x_2} & \cdots & \dfrac{\partial y_1}{\partial x_m} \\ \vdots & & & \\ \dfrac{\partial y_m}{\partial x_1} & \dfrac{\partial y_m}{\partial x_2} & \cdots & \dfrac{\partial y_m}{\partial x_m} \end{bmatrix}. \qquad (1.2)$$

The determinant of $\mathfrak{M}$, written $|\mathfrak{M}|$, is called the *Jacobian* of the transformation (1.1). It is known that if the transformation (1.1) carries a particular point $(x_1^0, \ldots, x_m^0)$ into the point $(y_1^0, \ldots, y_m^0)$ and if the Jacobian does not vanish at $(x_1^0, \ldots, x_m^0)$, then the Eqs. (1.1) allow a unique solution for the x_k in terms of the y_k for all points $y_1, \ldots, y_m$ sufficiently close to $y_1^0, \ldots, y_m^0$. Write it

$$x_k = x_k(y_1, \ldots, y_m), \qquad k = 1, \ldots, m. \qquad (1.3)$$

The partial derivatives $\partial x_k / \partial y_t$ are continuous in a neighborhood of $y_1^0, \ldots, y_m^0$. The matrix of the transformation (1.3) is the inverse* of the matrix $\mathfrak{M}$.

If n is an integer, the identity matrix I_n is the $n \times n$ matrix consisting of *ones* along the main diagonal and *zeros* elsewhere. By J or J_{2n}, we shall mean a certain matrix constructed in four blocks from I_n, namely,

$$J = J_{2n} = \begin{pmatrix} O_n & I_n \\ -I_n & O_n \end{pmatrix}, \qquad (1.4)$$

where O_n is the $n \times n$ matrix whose entries are all zero. It is easily verified that

$$J_{2n}^2 = -I_{2n}, \qquad J_{2n} = -J_{2n}^{-1}. \qquad (1.5)$$

Since $|J_{2n}^2| = |J_{2n}|^2 = |I_{2n}| = 1$, it follows that $|J_{2n}| \neq 0$.

Now let M_{2n} (we write M for simplicity) denote a $2n \times 2n$ matrix. It is called *symplectic* if

$$M^T J M = J, \qquad (1.6)$$

where M^T is the transpose of M. Since $|M^T| \cdot |J| \cdot |M| = |J|$ and $|M^T| = |M|$, then the non-vanishing of $|J|$ im-

---

* It is assumed that the reader is familiar with the notions of inverse and transpose of a matrix, and knows how to multiply matrices.

plies that $|M|^2 = 1$, $|M| = \pm 1$. Therefore M has an inverse M^{-1}, and from (1.5) and (1.6) we obtain

$$M^{-1} = -JM^TJ. \qquad (1.7)$$

A transformation (1.1) is called *canonical* if the corresponding Jacobian matrix $\mathfrak{M}$, defined by (1.2), is symplectic. Clearly, for such a transformation m must be even, $m = 2n$. In that case, it is customary to split the variables into p's and q's and write (1.1) in the form

$$p_k = p_k(P_1, \ldots, P_n; \quad Q_1, \ldots, Q_n), \qquad k = 1, \ldots, n,$$
$$q_k = q_k(P_1, \ldots, P_n; \quad Q_1, \ldots, Q_n), \qquad (1.8)$$

while the inverse transformation (1.3) takes the form

$$P_k = P_k(p_1, \ldots, p_n; \quad q_1, \ldots, q_n), \qquad k = 1, \ldots, n,$$
$$Q_k = Q_k(p_1, \ldots, p_n; \quad q_1, \ldots, q_n). \qquad (1.9)$$

EXERCISE 1.1.   Show that J is symplectic. Conclude that the transformation

$$p_k = Q_k; \quad q_k = -P_k, \qquad k = 1, \ldots, n$$

is canonical.

EXERCISE 1.2.   Use (1.4) to verify (1.5).

EXERCISE 1.3.   Give the details which establish (1.7).

*EXERCISE 1.4.   Let α and β denote the n $\times$ 1 matrices

$$\alpha = \begin{bmatrix} \alpha_1 \\ \vdots \\ \alpha_n \end{bmatrix}, \qquad \beta = \begin{bmatrix} \beta_1 \\ \vdots \\ \beta_n \end{bmatrix}.$$

Show that

$$J\begin{pmatrix} \alpha \\ \beta \end{pmatrix} = \begin{pmatrix} \beta \\ -\alpha \end{pmatrix}.$$

*EXERCISE 1.5.   Show that the transformation

$$p_1 = P_1\cos Q_2 - P_2 Q_1^{-1}\sin Q_2$$

$$p_2 = P_1\sin Q_2 + P_2 Q_1^{-1}\cos Q_2$$

$$q_1 = Q_1\cos Q_2$$

$$q_2 = Q_1\sin Q_2$$

is canonical. Suggestion: perform the multiplication $\mathcal{M}^T J \mathcal{M}$ in blocks of four, using the fact that the transpose of

$$\begin{pmatrix} A & B \\ C & D \end{pmatrix}$$

is

$$\begin{pmatrix} A^T & C^T \\ B^T & D^T \end{pmatrix}.$$

EXERCISE 1.6.   Prove that if M is symplectic, so is M^{-1}. If M_1 and M_2, each of order $2n \times 2n$, are symplectic, so is $M_1 M_2$.

Conclusion: the symplectic matrices of a fixed size form a group.

EXERCISE 1.7.   Interpret the preceding exercise when the matrices are Jacobian matrices of transformations.

EXERCISE 1.8.   (For matrix experts.) We know that if M is symplectic, then $|M| = \pm 1$. Prove that actually $|M| = +1$.

*EXERCISE 1.9.   Let $M = M_{2n}$ represent the matrix

$$\begin{pmatrix} A & B \\ C & D \end{pmatrix},$$

where each entry is an $n \times n$ matrix. Use the suggestion of Ex. 1.5 to evaluate $M^T J M$. Show from this that M is symplectic if and only if
   (i) $A^T C$, $B^T D$ are symmetric (that is, are their own transposes);
   (ii) $D^T A - B^T C = I$.

*EXERCISE 1.10.   Apply the preceding exercise to the matrix $\mathfrak{M}$ of the transformation (1.8). Conclude that it is symplectic and the transformation canonical if and only if

$$\sum_{m=1}^{n} \left[\frac{\partial p_m}{\partial P_k} \frac{\partial q_m}{\partial P_l} - \frac{\partial p_m}{\partial P_l} \frac{\partial q_m}{\partial P_k} \right] = 0$$

$$\sum_{m=1}^{n} \left[\frac{\partial p_m}{\partial Q_k} \frac{\partial q_m}{\partial Q_l} - \frac{\partial p_m}{\partial Q_l} \frac{\partial q_m}{\partial Q_k} \right] = 0$$

for all k, l and

$$\sum_{m=1}^{n} \left[\frac{\partial p_m}{\partial P_k} \frac{\partial q_m}{\partial Q_l} - \frac{\partial p_m}{\partial Q_l} \frac{\partial q_m}{\partial P_k} \right] = \delta_{kl}.$$

The symbol δ_{kl} means 1 when $k = l$, and 0 when $k \neq l$.

*EXERCISE 1.11.   Let (1.8) be a given transformation. Then

$$dq_k = \sum_{l=1}^{n} \left[\frac{\partial q_k}{\partial P_l} dP_l + \frac{\partial q_k}{\partial Q_l} dQ_l \right].$$

In the expression

$$\sum_{k=1}^{n} p_k dq_k - P_k dQ_k$$

replace p_k as given by (1.8) and dq_k as just obtained. The result, after rearrangement, is the differential form K, defined by

$$K = \sum_{l=1}^{n} A_l dP_l + B_l dQ_l,$$

where A_l and B_l are functions only of P_k and Q_k. Prove, by use of Ex. 1.10 that (1.8) defines a canonical transformation if and only if there is a function $W(P_1, \ldots, P_n; Q_1, \ldots, Q_n)$ whose total differential is K.

*EXERCISE 1.12.   Show by the method described in Ex. 1.11 that the transformation $p_1 = P_1, p_2 = Q_2, q_1 = Q_1, q_2 = -P_2$ is canonical. Do the same for the (Legendre) transformation described in Ex. 1.1 and for the transformation of Ex. 1.5.

2. AN APPLICATION OF CANONICAL TRANSFORMATIONS

We have seen on several occasions that the equations of a system may be put in the form

$$\dot{q}_k = \frac{\partial H}{\partial p_k}, \quad \dot{p}_k = -\frac{\partial H}{\partial q_k}, \quad k = 1, \ldots, m, \quad (2.1)$$

where the function H, the *Hamiltonian* of the system, is a function of $p_1, \ldots, p_m; q_1, \ldots, q_m$. Let

$$p_k = p_k(P_1, \ldots, P_m; Q_1, \ldots, Q_m)$$
$$q_k = q_k(P_1, \ldots, P_m; Q_1, \ldots, Q_m) \tag{2.2}$$

represent a canonical transformation. With this replacement of the original variables, H becomes a function of $P_1, \ldots, P_m; Q_1, \ldots, Q_m$. We shall show that *the system* (2.1) *retains its original form under this transformation*, that is,

$$\dot{Q}_k = \frac{\partial H}{\partial P_k}, \quad \dot{P}_k = -\frac{\partial H}{\partial Q_k}, \quad k = 1, \ldots, m. \tag{2.3}$$

For ease of writing, we adopt the notation $\dot{p}$ for the $n \times 1$ matrix with entries $\dot{p}_1, \dot{p}_2, \ldots, \dot{p}_n$, and similarly for $\dot{q}, \dot{P}, \dot{Q}$. The functions of system (2.2) have the derivatives

$$\dot{p}_k = \sum_{l=1}^{m} \left[\frac{\partial p_k}{\partial P_l} \dot{P}_l + \frac{\partial p_k}{\partial Q_l} \dot{Q}_l \right]$$

$$\dot{q}_k = \sum_{l=1}^{m} \left[\frac{\partial q_k}{\partial P_l} \dot{P}_l + \frac{\partial q_k}{\partial Q_l} \dot{Q}_l \right].$$

If $\mathfrak{M}$ is the Jacobian matrix of (2.2), this says that

$$\begin{pmatrix} \dot{p} \\ \dot{q} \end{pmatrix} = \mathfrak{M} \begin{pmatrix} \dot{P} \\ \dot{Q} \end{pmatrix}.$$

Therefore

$$\begin{pmatrix} \dot{P} \\ \dot{Q} \end{pmatrix} = \mathfrak{M}^{-1} \begin{pmatrix} \dot{p} \\ \dot{q} \end{pmatrix}.$$

According to (1.7), this is the same as

$$\begin{pmatrix} \dot{P} \\ \dot{Q} \end{pmatrix} = -J \mathfrak{M}^T J \begin{pmatrix} \dot{p} \\ \dot{q} \end{pmatrix},$$

because $\mathcal{M}$ is symplectic. By Ex. 1.4, this says that

$$\begin{pmatrix} \dot{P} \\ \dot{Q} \end{pmatrix} = -J\mathcal{M}^T \begin{pmatrix} \dot{q} \\ -\dot{p} \end{pmatrix}.$$

Left-multiply each side by J. Since $J^2 = -I$

$$\begin{pmatrix} \dot{Q} \\ -\dot{P} \end{pmatrix} = \mathcal{M}^T \begin{pmatrix} \dot{q} \\ -\dot{p} \end{pmatrix}.$$

By (2.1), this becomes

$$\begin{pmatrix} \dot{Q} \\ -\dot{P} \end{pmatrix} = \mathcal{M}^T \begin{pmatrix} \dfrac{\partial H}{\partial p} \\[2mm] \dfrac{\partial H}{\partial q} \end{pmatrix}.$$

Multiplication of the two matrices on the right-hand side shows that

$$\dot{Q}_k = \sum_{l=1}^{m} \left[\frac{\partial H}{\partial p_l}\frac{\partial p_l}{\partial P_k} + \frac{\partial H}{\partial q_l}\frac{\partial q_l}{\partial P_k} \right]$$

$$\dot{P}_k = \sum_{l=1}^{m} \left[\frac{\partial H}{\partial p_l}\frac{\partial p_l}{\partial Q_k} + \frac{\partial H}{\partial q_l}\frac{\partial q_l}{\partial Q_k} \right]. \tag{2.4}$$

Finally, the chain rule for differentiation shows that the right-hand sides of (2.3) and (2.4) are identical. This completes the proof of the assertion that the Hamiltonian form (2.1) is preserved under a canonical transformation, with H undergoing the change of variables (2.2).

As an illustration, we turn to Ex. 11.2 of the preceding chapter, first making a notational change. Write p_1 for p, p_2 for P, q_1 for ξ, q_2 for η. The function H becomes

$$\tfrac{1}{2}(p_1 + q_2)^2 + \tfrac{1}{2}(p_2 - q_1)^2 - \Phi(q_1, q_2)$$

$$= \tfrac{1}{2}(p_1^2 + p_2^2) - (q_1 p_2 - q_2 p_1) + \tfrac{1}{2}(q_1^2 + q_2^2) - \Phi(q, q_2).$$

According to (11.1) of the preceding chapter, this is the same as

$$\tfrac{1}{2}\left(p_1^2 + p_2^2\right) - (q_1 p_2 - q_2 p_1) - U(q_1, q_2) - \tfrac{1}{2}\,\mu(1-\mu),$$

$$(2.5)$$

and the differential equations of the circular restricted problem take the form (2.1) with $m = 2$ and H defined by (2.5).

We now apply the canonical transformation of Ex. 1.5, namely

$$p_1 = P_1 \cos Q_2 - P_2 Q_1^{-1}\sin Q_2$$

$$p_2 = P_1 \sin Q_2 + P_2 Q_1^{-1}\cos Q_2$$

$$q_1 = Q_1 \cos Q_2$$

$$q_2 = Q_1 \sin Q_2.$$

The Hamiltonian (2.5) becomes

$$\tfrac{1}{2}\left(P_1^2 + P_2^2 Q_1^{-2}\right) - P_2 - U(Q_1\cos Q_2,\ Q_1\sin Q_2)$$

$$- \tfrac{1}{2}\,\mu(1-\mu) \qquad\qquad (2.6)$$

and the equations are (2.3) with $m = 2$.

EXERCISE 2.1. By retracing all the variables back to the original (non-rotating) system x-y, show that the terms of the Hamiltonian (2.6) of the restricted circular problem have these interpretations:

$$P_1^2 + P_2^2 Q_1^{-2} = v^2 = \dot{x}^2 + \dot{y}^2,$$

$$P_2 = c = x\dot{y} - y\dot{x}.$$

The quantity v is the velocity of the particle in the original coordinate system and c is its angular momentum. Observe also that Q_1, Q_2 represent the polar coordinates of the particle in the (rotating) ξ-η system. What do P_1, P_2 mean?

3. CANONICAL TRANSFORMATIONS GENERATED BY A FUNCTION

In this section the symbol $\sum$ means $\sum\limits_{k=1}^{m}$.

Let

$$p_k = p_k(P_1, \ldots, P_m; \quad Q_1, \ldots, Q_m)$$
$$q_k = q_k(P_1, \ldots, P_m; \quad Q_1, \ldots, Q_m), \tag{3.1}$$

$k = 1, \ldots, m$, denote a transformation. In the preceding section it was shown that *the transformation is canonical if and only if the differential form*

$$\sum p_k dq_k - P_k dQ_k, \tag{3.2}$$

after replacement of p_k and dq_k from (3.1), is exact in the P_k and Q_k. This form is related to three others:

$$\sum q_k dp_k + P_k dQ_k, \tag{3.3}$$

$$\sum p_k dq_k + Q_k dP_k, \tag{3.4}$$

$$\sum q_k dp_k - Q_k dP_k. \tag{3.5}$$

If we denote each of the four forms by F_i, $i = 1, 2, 3, 4$, respectively, it is easy to verify that

$$F_1 = -F_2 + d\sum p_k q_k,$$

$$F_1 = F_3 - d\sum P_k Q_k,$$

$$F_1 = -F_4 + d\sum (p_k q_k - P_k Q_k).$$

It follows that if any one of the four differential forms *after replacement* of p_k, q_k, dp_k, dq_k from (3.1) is exact, that is, the differential of a function of the P_k and Q_k, so is each of the others. Therefore the transformation (3.1) is canonical if and only if any *one* of the forms is exact after the replacement.

A subtlety, often overlooked, must be mentioned here. We illustrate with the form (3.3) and $m = 2$, although the comments apply in the other cases. To say that (3.3) is exact after replacement of q_k and dp_k does *not* mean that there is a function $S(p_1, p_2; Q_1, Q_2)$ whose differential

$$dS = \frac{\partial S}{\partial p_1} dp_1 + \frac{\partial S}{\partial p_2} dp_2 + \frac{\partial S}{\partial Q_1} dQ_1 + \frac{\partial S}{\partial Q_2} dQ_2$$

agrees with (3.3), namely,

$$q_1 dp_1 + q_2 dp_2 + P_1 dQ_1 + P_2 dQ_2,$$

in the sense that the relations

$$\frac{\partial S}{\partial p_1} = q_1, \qquad \frac{\partial S}{\partial p_2} = q_2, \qquad \frac{\partial S}{\partial Q_1} = P_1, \qquad \frac{\partial S}{\partial Q_2} = P_2$$

$$(3.6)$$

hold identically after the replacement. For example, let $p_1 = P_1$, $p_2 = Q_2$, $q_1 = Q_1$, $q_2 = -P_2$ be a transformation; according to Ex. 1.11, it is canonical. The form (3.3), on replacement of p_k and dq_k, becomes

$$Q_1 dP_1 - P_2 dQ_2 + P_1 dQ_1 + P_2 dQ_2,$$

which is $d(P_1 Q_1)$ and exact. But there is no function $S(p_1, p_2; Q_1, Q_2)$ for which (3.6) is satisfied. To show this, look at the second equation in (3.6). It says

$$\frac{\partial S}{\partial p_2}(p_1, p_2; Q_1, Q_2) = q_2,$$

or

$$\frac{\partial S}{\partial p_2}(P_1, Q_2; Q_1, Q_2) = -P_2,$$

which is impossible since the left-hand side does not contain P_2.

On the other hand, it *may* happen for some canonical transformation that there *is* a function $S(p_1, p_2; Q_1, Q_2)$ for which (3.6) is satisfied. Consider, for example, the canonical transformation of Ex. 1.5, namely,

$$
\begin{aligned}
p_1 &= P_1 \cos Q_2 - P_2 Q_1^{-1} \sin Q_2 \\
p_2 &= P_1 \sin Q_2 + P_2 Q_1^{-1} \cos Q_2 \\
q_1 &= Q_1 \cos Q_2 \\
q_2 &= Q_1 \sin Q_2.
\end{aligned}
\tag{3.7}
$$

We ask whether there is a function $S(p_1, p_2; Q_1, Q_2)$

satisfying (3.6) identically. The first two equations of (3.6) read

$$\frac{\partial S}{\partial p_1} = Q_1 \cos Q_2$$

$$\frac{\partial S}{\partial p_2} = Q_1 \sin Q_2.$$

It follows that an admissible S must be of the form $p_1 Q_1 \cos Q_2 + p_2 Q_1 \sin Q_2 + T$, where T is a function of Q_1 and Q_2 only. The last two equations of (3.6) then require that

$$p_1 \cos Q_2 + p_2 \sin Q_2 + \frac{\partial T}{\partial Q_1} = P_1$$

$$-p_1 Q_1 \sin Q_2 + p_2 Q_1 \cos Q_2 + \frac{\partial T}{\partial Q_2} = P_2.$$

Substituting for p_1 and p_2 from (3.7), we obtain $\partial T/\partial Q_1 = 0$, $\partial T/\partial Q_2 = 0$, so that T is a constant. Since only the derivatives of S appear in (3.6), we can drop the constant to conclude that $S(p_1, p_2; Q_1, Q_2)$, defined as $Q_1(p_1 \cos Q_2 + p_2 \sin Q_2)$, accomplishes the desired purpose.

If a function S of the desired form does exist satisfying (3.6), we call it a *generating* function for the canonical transformation. We have concentrated on the form (3.3), but analogous results hold for the other forms.

EXERCISE 3.1. Show that the transformation $p = P \cos Q$, $q = P \sin Q$ (where $m = 1$) is not canonical, but that its modification $p = \sqrt{2P} \cos Q$, $q = \sqrt{2P} \sin Q$ is. Find a generating function $S(q, Q)$.

EXERCISE 3.2.   Discuss the two canonical transformations described in this section by replacing (3.3) in turn by each of the other three forms (3.2), (3.4), (3.5) and the Eqs. (3.6) in turn by the correct analogues. Show, in particular, that the first transformation does not have a generating function in any of the four arrangements. How about the second transformation?

EXERCISE 3.3.   Are there other generating functions for the transformation of Ex. 3.1?

4.   GENERATING FUNCTIONS

Let

$$p_k = p_k(P_1, \ldots, P_m; Q_1, \ldots, Q_m)$$
$$q_k = q_k(P_1, \ldots, P_m; Q_1, \ldots, Q_m) \tag{4.1}$$

denote a canonical transformation.

Then

(i)   there is a function $S(q_1, \ldots, q_m; Q_1, \ldots, Q_m)$ whose differential is (3.2), that is, for which

$$\frac{\partial S}{\partial q_k} = p_k, \qquad \frac{\partial S}{\partial Q_k} = -P_k; \tag{4.2}$$

or

(ii)   there is a function $S(p_1, \ldots, p_m; Q_1, \ldots, Q_m)$ whose differential is (3.3), that is, for which

$$\frac{\partial S}{\partial p_k} = q_k, \qquad \frac{\partial S}{\partial Q_k} = P_k; \tag{4.3}$$

or

(iii)   there is a function $S(q_1, \ldots, q_m; P_1, \ldots, P_m)$ .whose differential is (3.4), that is, for which

$$\frac{\partial S}{\partial q_k} = p_k, \qquad \frac{\partial S}{\partial P_k} = Q_k; \qquad (4.4)$$

or

(iv)   there is a function $S(p_1, \ldots, p_m; P_1, \ldots, P_m)$ whose differential is (3.5), that is, for which

$$\frac{\partial S}{\partial p_k} = q_k, \qquad \frac{\partial S}{\partial P_k} = -Q_k; \qquad (4.5)$$

or

(v)   none of (i), (ii), (iii), (iv) is true.

We have seen (Ex. 3.2) that case (v) can actually occur.

Now forget the transformation (4.1). Suppose we *start* with a function S of one of the four forms described in (i)–(iv). For definiteness, let us say S is of the form (iii). Let us *define* the variables p_k, Q_k by (4.4). The second of these equations is

$$\frac{\partial S}{\partial P_k} (q_1, \ldots, q_m; P_1, \ldots, P_m) = Q_k, \qquad k = 1, \ldots, m.$$

Suppose, moreover, that the *Hessian* $|\partial^2 S / \partial P_k \partial q_l|$ does not vanish. Then this system of m equations can be solved for the q_k in terms of the Q_k and P_k yielding functions of the form (4.1). The first of the Eqs. (4.4) can be written

$$p_k = \frac{\partial S}{\partial q_k} (q_1, \ldots, q_m; P_1, \ldots, P_m).$$

Replacing the q_k by $q_k(P_1, \ldots, P_m; Q_1, \ldots, Q_m)$ yields for p_k of the form (4.1). Clearly, the transformation (4.1)

so obtained is generated by the function S, and is, therefore, canonical.

The same argument can be applied to the three other
forms of S. The technique provides a method for obtain-
ing canonical transformations, *starting* with a function S.
The implications are very important, as we now show.

Suppose that we are given a system of differential
equations

$$\dot{q}_k = \frac{\partial H}{\partial p_k} , \quad \dot{p}_k = -\frac{\partial H}{\partial q_k} , \qquad k = 1, \ldots, m \quad (4.6)$$

with Hamiltonian $H(p_1, \ldots, p_m; q_1, \ldots, q_m)$. Then it retains its form under a canonical transformation (4.1); that
is, after the change of variables in the Hamiltonian, the
system becomes

$$\dot{Q}_k = \frac{\partial H}{\partial P_k} , \quad \dot{P}_k = -\frac{\partial H}{\partial Q_k} , \qquad k = 1, \ldots, m. \quad (4.7)$$

Now let us try to find a canonical transformation which
reduces H to a very simple form so that the system (4.7) is
manageable. For example, suppose it is possible to find a
substitution (4.1) such that H reduces identically to Q_1.
Then Eqs. (4.7) become

$$\dot{Q}_k = 0, \qquad k = 1, \ldots, m;$$

$$\dot{P}_1 = -1, \quad \dot{P}_k = 0, \qquad k = 2, \ldots, m.$$

Then

$$P_1 = -t + \alpha_1, \quad P_k = \alpha_k, \quad k = 2, \ldots, m;$$
$$Q_k = \beta_k, \qquad\qquad k = 1, \ldots, m, \qquad (4.8)$$

where all the α_k, β_k are constants. Substitution for P_k, Q_k

into (4.1) then gives the solution of (4.6) in terms of t and the "arbitrary" constants $\alpha_1, \ldots, \alpha_m; \beta_1, \ldots, \beta_m$.

But how can one find a transformation (4.1) which does in fact reduce H to Q_1? A procedure, due to Jacobi, is to search for a generating function S that produces such a transformation. Specifically, let us try for a function S of type (i). If the p_k in $H(p_1, \ldots, p_m; q_1, \ldots, q_m)$ are replaced from (4.2), we get

$$H\left(\frac{\partial S}{\partial q_1}, \ldots, \frac{\partial S}{\partial q_m} ; q_1, \ldots, q_m \right),$$

where S is of the form $S(q_1, \ldots, q_m; Q_1, \ldots, Q_m)$ and we are asking that

$$H\left(\frac{\partial S}{\partial q_1}, \ldots, \frac{\partial S}{\partial q_m}\; q_1, \ldots, q_m \right) = Q_1, \qquad (4.9)$$

irrespective of the values of $Q_2, \ldots, Q_m$. This is the *Jacobi* (partial differential) *equation* . If we can find such an S and the Hessian $|\partial^2 S/\partial q_k \partial Q_l|$ does not vanish, then S generates a transformation (4.1) which has the desired properties.

A fairly simple example may help to make all this clearer. Let $m = 1$ (so that we need no subscripts) and let $H = \frac{1}{2}(p^2 + q^2)$. The differential equations are $\dot{q} = \partial H/\partial p = p$, $\dot{p} = -\partial H/\partial q = -q$. They are trivial to solve, since $\ddot{p} + p = 0$, $p = A \cos(t - B)$, $q = -\dot{p} = A \sin(t - B)$. But we wish to solve them by the method outlined above, because direct integration of a system is seldom possible.

We seek $S(q, Q)$ so that (4.9), in this case

$$\frac{1}{2}\left[\left(\frac{\partial S}{\partial q} \right)^2 + q^2 \right] = Q,$$

is satisfied. Then $\partial S/\partial q = (2Q - q^2)^{1/2}$ and

$$S = \tfrac{1}{2}\left[q(2Q - q^2)^{1/2} + 2Q \text{ arc sin } q(2Q)^{-1/2} \right].$$

Therefore

$$-P = \frac{\partial S}{\partial Q} = \text{arc sin } q(2Q)^{-1/2}$$

$$p = \frac{\partial S}{\partial q} = (2Q - q^2)^{1/2}.$$

According to (4.8), $P = -t + \alpha$, $Q = \beta$. Therefore $q = \sqrt{2\beta} \sin(t - \alpha)$, $p = \sqrt{2\beta} \cos(t - \alpha)$, which certainly provides a general solution of the equation.

EXERCISE 4.1.   What is the Hessian in the example just worked?

EXERCISE 4.2.   In the general case can the reduction of H to Q_1 be accomplished by an S of one of the other three types? Suppose we had tried for a reduction to P_1. What modifications are needed? Try out your theory on the special example.

*EXERCISE 4.3.   Show that if S satisfies Eq. (4.9), then the solution p_k, q_k of (4.6) is given "implicitly" by

$$-t + \alpha_1 = -\frac{\partial S}{\partial \beta_1}(q_1, \ldots, q_m; \beta_1, \ldots, \beta_m),$$

$$\alpha_k = -\frac{\partial S}{\partial \beta_k}(q_1, \ldots, q_m; \beta_1, \ldots, \beta_m), \quad k = 2, \ldots, m,$$

$$p_k = \frac{\partial S}{\partial q_k}(q_1, \ldots, q_m; \beta_1, \ldots, \beta_m), \quad k = 1, \ldots, m,$$

where the $\alpha_1, \ldots, \alpha_m; \beta_1, \ldots, \beta_m$ are $2m$ arbitrary constants.

EXERCISE 4.4.   What happens in the example of the text if we choose $\partial S/\partial q = -(2Q - q^2)^{1/2}$ instead of the positive square root?

5.   APPLICATION TO THE CENTRAL FORCE AND RESTRICTED PROBLEMS

We have seen in Sec. 2 that the restricted three-body problem can be put in the form

$$\dot{q}_k = \frac{\partial H}{\partial p_k}, \quad \dot{p}_k = -\frac{\partial H}{\partial q_k}, \quad k = 1, 2, \qquad (5.1)$$

with

$$H = \tfrac{1}{2}\left(p_1^2 + p_2^2 q_1^{-2}\right) - p_2 - U(q_1\cos q_2, q_1\sin q_2).$$

Observe that we have changed from capital letters to small; this is because another canonical transformation is forthcoming. The constant $\tfrac{1}{2}\mu(1 - \mu)$ has been dropped from the Hamiltonian; no harm is done since it does not appear in the Eqs. (5.1) anyhow.

Recall that, according to Ex. 2.1, the term $\tfrac{1}{2}(p_1^2 + p_2^2 q_1^{-2})$ is simply $\tfrac{1}{2}v^2$, where v is the velocity of the particle in the *non-rotating* system and that $p_2 = c$, where c is the angular momentum. The variables q_1, q_2 are the polar coordinates of the particle in the *rotating* coordinate system. In the latter system, U is defined by

$$U(\xi, \eta) = (1 - \mu)\left[(\xi + \mu)^2 + \eta^2\right]^{-1/2}$$
$$+ \mu\left[(\xi + \mu - 1)^2 + \eta^2\right]^{-1/2},$$

in accordance with the formula of Ex. 10.3 of Chap. 1. In the special case $\mu = 0$, the function $U(\xi, \eta)$ becomes simply $(\xi^2 + \eta^2)^{-1/2}$, which is the same as q_1^{-1}. This suggests rewriting the Hamiltonian as

$$H = H_0 + \left[q_1^{-1} - U \right], \qquad (5.2)$$

where

$$H_0 = \tfrac{1}{2}\left(p_1^2 + p_2^2 q_1^2 \right) - q_1^{-1} - p_2, \qquad (5.3)$$

and where the term in brackets drops out when $\mu = 0$.

What is the physical meaning of the problem if $\mu = 0$? The answer is simple: the smaller primary mass disappears and the large one takes on the total mass of unity at the origin. The problem is then that of a mass moving in a fixed plane under the attraction of a central force. Since the mass at O is unity, this is identical with the problem $\ddot{\mathbf{r}} = -r^{-3}\mathbf{r}$ treated in Chap. 1. In the current context, the problem takes the form (5.1), with H replaced by H_0. Therefore, the central force problem can be put in the form

$$\dot{q}_k = \frac{\partial H_0}{\partial p_k}, \qquad \dot{p}_k = -\frac{\partial H_0}{\partial q_k} \qquad (5.4)$$

where H_0 is defined by (5.3). According to the second paragraph of this section, H_0 can be written $(\tfrac{1}{2}v^2 - r^{-1}) - c$, where $r = q_1$. The first term is just the energy h of the moving particle. Therefore $H_0 = h - c$. This suggests a new canonical transformation to simplify H_0, and hence the Eqs. (5.4). It is reasonable to let $Q_1 = h$, $Q_2 = c$, so that $H_0 = Q_1 - Q_2$. If we can find such a transformation, the Eqs. (5.4) will become

$$\dot{Q}_k = \frac{\partial H_0}{\partial P_k}, \qquad \dot{P}_k = -\frac{\partial H_0}{\partial Q_k}, \qquad k = 1, 2,$$

so that

$$\dot{Q}_1 = 0, \qquad \dot{Q}_2 = 0, \qquad \dot{P}_1 = -1, \qquad \dot{P}_2 = 1, \quad (5.5)$$

which are easy to solve.

With this in mind, let

$$Q_1 = \tfrac{1}{2}\left(p_1^2 + p_2^2 q_1^{-2}\right) - q_1^{-1}$$
$$Q_2 = p_2. \tag{5.6}$$

But how are P_1, P_2 to be chosen so that the transformation is canonical? Let us look for a generating function that will furnish the desired transformation. Since it is P_1, P_2 that are missing, we shall try for a function of the form $S(q_1, q_2; Q_1, Q_2)$, for then $-P_k = \partial S/\partial Q_k$, according to (4.2). Since $p_2 = \partial S/\partial q_2$, the second of Eqs. (5.6) requires that $\partial S/\partial q_2 = Q_2$. Therefore $S = q_2 Q_2 + F$, where F cannot depend on q_2 and must be of the form $F(q_1; Q_1, Q_2)$. Since $p_1 = \partial S/\partial q_1 = \partial F/\partial q_1$, the first of Eqs. (5.6) demands that

$$\tfrac{1}{2}\left[\left(\frac{\partial F}{\partial q_1}\right)^2 + Q_2^2 q_1^{-2}\right] - q_1^{-1} = Q_1.$$

Since any solution will serve our purpose, let

$$\frac{\partial F}{\partial q_1} = q_1^{-1}\left(-Q_2^2 + 2q_1 + 2Q_1 q_1^2\right)^{1/2}.$$

Therefore

$$F = \int_G^{q_1} x^{-1}\left(-Q_2^2 + 2x + 2Q_1 x^2\right)^{1/2} dx,$$

where G depends only on Q_1 and Q_2. As we shall see shortly, it is best to choose G so that the integrand vanishes when $x = G$; and because $q_1 > 0$, it is best to choose G positive.

If $Q_1 = h = 0$, there is just one choice for G, namely $Q = Q_2^2/2 = c^2/2$. If $Q_1 \neq 0$, we must choose between

$$\left[-1 \pm \left(1 + 2Q_1Q_2^2\right)^{1/2}\right](2Q_1)^{-1}.$$

The term $1 + 2Q_1Q_2^2$ is $1 + 2hc^2$. According to equation (5.2) of Chap. 1 (with μ as defined there equal to unity), this is the same as e^2. Therefore $G = (-1 \pm e)(2h)^{-1}$. If $h > 0$, we are forced to choose the $+$ sign to make G positive. If $h < 0$, either sign makes G positive but we must choose the larger value to keep the integrand real for $x > G$. In either case, then, $G = (-1 + e)(2h)^{-1}$. In summary, we have chosen for S the function $q_2Q_2 + F$, òr

$$S(q_1, q_2; Q_1, Q_2) = q_2Q_2$$
$$+ \int_G^{q_1} x^{-1}\left(-Q_2^2 + 2x + 2Q_1x^2\right)^{1/2}dx. \quad (5.7)$$

The "missing" variables P_1, P_2 are then defined by $-\partial S/\partial Q_1$ and $-\partial S/\partial Q_2$, respectively. What is their physical interpretation? We start with P_2. According to (5.7) and Leibniz' rule for differentiation of an integral,

$$-P_2 = q_2 - Q_2\int_G^{q_1} x^{-1}\left(-Q_2^2 + 2x + 2Q_1x^2\right)^{-1/2}dx$$
$$\tag{5.8}$$
$$= q_2 + \text{arc cos } \frac{q_1 - Q_2^2}{q_1e} - \text{arc cos } \frac{G - Q_2^2}{Ge}.$$

This step uses the fact that the integrand of (5.7) vanishes at $x = G$. Now let $q_1 = r = c^2(1 + e \cos f)^{-1}$, $Q_2^2 = c^2$. The second term on the right-hand side of (5.8) becomes arc cos $(-\cos f) = \pi - f$. Therefore,

$$-P_2 = (q_2 - f) + \left(\pi - \text{arc cos} \frac{G - Q_2^2}{Ge}\right).$$

By our choice of G, the last term vanishes. Therefore $-P_2 = q_2 - f$. Recall that q_2 is the angle made by the radius vector to the particle with the positive ξ-axis. It follows that $q_2 - f$ is the amplitude of pericenter, *measured from the ξ-axis*. As a check, observe that at time t the ξ-axis forms an angle of t with the fixed x-axis, since the rotation rate is 1. Therefore $t + (q_2 - f)$ is ω, the (constant) amplitude of pericenter measured from the x-axis. We conclude that $-P_2 = \omega - t$, $\dot{P}_2 = 1$, which is consistent with (5.5).

It remains to interpret P_1, which is defined by

$$-P_1 = \frac{\partial S}{\partial Q_1} = Q_1 \int_G^{q_1} x \left(-Q_2^2 + 2x + 2Q_1 x^2\right)^{-1/2} dx.$$

$$(5.9)$$

The interpretation is left to the exercises which follow.

*EXERCISE 5.1.   Assume elliptic motion, that is, $Q_1 = h < 0$. Carry out the integration in (5.9) to obtain

$$-P_1 = (-2Q_1)^{-1} \Bigg[-\left(-Q_2^2 + 2q_1 + 2Q_1 q_1^2\right)^{1/2}$$

$$+ (-2Q_2)^{-1/2} \text{arc cos} \frac{2Q_1 q_1 + 1}{e} \Bigg].$$

*EXERCISE 5.2. In the preceding formula, let $Q_1 = h$, $a = -1/2h$, $q_1 = r = a(1 - e \cos u)$, $Q_2 = c$, where u is the eccentric anomaly. Show that

$$-P_1 = a^{3/2}(u - e \sin u).$$

Finally, let $n = a^{-3/2}$ and conclude that $-P_1 = t - T$, where T is time of pericenter passage. Observe that $\dot{P}_1 = -1$, again consistent with (5.5).

EXERCISE 5.3. By use of Eqs. (7.5) of Chap. 1 (with μ there set equal to unity) verify that the interpretation $-P_1 = t - T$ is valid in case $h = 0$, $e = 1$.

EXERCISE 5.4. Show that $-P_1 = t - T$ is also valid when $h > 0$.

EXERCISE 5.5. Show that in terms of the "old" variables p_1, p_2; q_1, q_2, we have

$$e^2 = (p_1 p_2)^2 + \left(1 - p_2^2 q_1^{-1}\right)^2$$

and

$$-P_2 = q_2 - \text{arc} \cos \frac{p_2^2 - q_1}{q_1 e}.$$

Show that $-P_1$ can also be expressed explicitly in terms of the "old" variables, but do not write out the expression.

EXERCISE 5.6. Show that G is simply the distance of the mass from O at pericenter passage.

*EXERCISE 5.7. Apply the transformation of this sec tion to the *original* Eqs. (5.1). They become

$$\dot{Q}_k = \frac{\partial H}{\partial P_k}, \quad \dot{P}_k = -\frac{\partial H}{\partial Q_k}, \quad k = 1, 2,$$

where

$$H = Q_1 - Q_2 + R,$$

R being the result of substituting the new variables into $[q_1^{-1} - U]$.

Reinterpret the restricted problem as a problem of central force motion with a disturbance represented by the term R in the Hamiltonian. The constants c, h, ω, T now become functions of time. In particular, since $Q_2 = c$, we get $\dot{c} = \partial H/\partial P_2 = \partial R/\partial P_2$. Prove that this agrees with the formula $\dot{c} = rF_\alpha$ appearing in (17.8), Chap. 1.

6.   EQUILIBRIUM POINTS AND THEIR STABILITY

Let a system be governed by the equations

$$\dot{q}_k = \frac{\partial H}{\partial p_k}, \quad \dot{p}_k = -\frac{\partial H}{\partial q_k}, \quad k = 1, \ldots, m, \quad (6.1)$$

with Hamiltonian $H(p_1, \ldots, p_m; q_1, \ldots, q_m)$. Suppose that $p_1^0, \ldots, p_m^0; q_1^0, \ldots, q_m^0$ is a point at which all the first partial derivatives of H vanish; it is called an *equilibrium point*. Then the set of constant functions $p_k = p_k^0$, $q_k = q_k^0$ satisfy the differential equations; it is called an *equilibrium solution*.

The major example for our purposes occurs in the restricted three-body problem. Consider the problem in the form (6.1) with $m = 2$ and the Hamiltonian given by (2.5), namely,

$$\tfrac{1}{2}\left(p_1^2 + p_2^2\right) - (q_1 p_2 - q_2 p_1) - U(q_1, q_2) - \tfrac{1}{2}\mu(1 - \mu).$$

$$(6.2)$$

The four partial derivatives are

$$\frac{\partial H}{\partial p_1} = p_1 + q_2, \qquad \frac{\partial H}{\partial q_1} = -p_2 - \frac{\partial U}{\partial q_1},$$

$$\frac{\partial H}{\partial p_2} = p_2 - q_1, \qquad \frac{\partial H}{\partial q_2} = p_1 - \frac{\partial U}{\partial q_2}. \tag{6.3}$$

Clearly, they all vanish at a point $(p_1, p_2; q_1, q_2)$ if and only if $p_1 = -q_2$, $p_2 = q_1$, $q_1 + (\partial U/\partial q_1) = 0$, $q_2 + (\partial U/\partial q_2) = 0$. Since $q_1 = \xi$, $q_2 = \eta$, the last two equations are identical with (12.1) of Chap. 2. Hence the equilibrium points for this Hamiltonian system are the five points $(-q_2^0, q_1^0; q_1^0, q_2^0)$, where (q_1^0, q_2^0) is any one of the five libration points.

We return to the general problem (6.1) and an equilibrium point $(p_1^0, \ldots, p_m^0; q_1^0, \ldots, q_m^0)$. An important question concerning such a point is this: will a "small" disturbance in the coordinates of this point cause the resulting solution of the system to depart considerably from the point? It is customary to call the point *stable* if the following is true: if a solution of (6.1) starts with initial conditions sufficiently "near" $(p_1^0, \ldots, p_m^0; q_1^0, \ldots, q_m^0)$, it will remain near this position for all time. In the special problem of the libration points we are asking: if a particle is placed near one of the libration points with (relative) velocity near zero, will it remain near this position for all time? Since $\frac{1}{2}(\dot{\xi}^2 + \dot{\eta}^2) + \Phi(\xi, \eta)$ is constant, this means that the velocities must also remain small for all time.

The general question can be put in a more precise form in terms of the concept of L-stability.* To explain what this means, let the distance between two points

---

* L for Liapounov.

$(p_1, \ldots, p_m; q_1, \ldots, q_m)$ and $(p'_1, \ldots, p'_m; q'_1, \ldots, q'_m)$ be measured by

$$\left[\Sigma (p_k - p'_k)^2 + (q_k - q'_k)^2 \right]^{1/2}.$$

Then a point of equilibrium is called L-*stable* if, for each positive number ϵ, there is a positive number δ such that each solution of (6.1) which starts with initial position within a distance δ of the point exists for all time *thereafter* and never departs from this point to a distance exceeding ϵ. Clearly, if there is such a δ, it must satisfy $\delta \leqslant \epsilon$.

The problem of L-stability of the libration points is a very difficult one and will be discussed in the sequel. Here we shall describe a simpler example, due to T. Cherry. Let $m = 2$ and let $H(p_1, p_2; q_1, q_2)$ be the Hamiltonian

$$\tfrac{1}{2}\left(q_1^2 + p_1^2\right) - \left(q_2^2 + p_2^2\right) + \tfrac{1}{2}\left(p_1^2 p_2 - p_2 q_1^2 - 2q_1 q_2 p_1\right). \quad (6.4)$$

Then the Eqs. (6.1) become

$$\begin{aligned}
\dot{p}_1 &= -q_1 + p_2 q_1 + q_2 p_1, \\
\dot{p}_2 &= 2q_2 + p_1 q_1, \\
\dot{q}_1 &= p_1 + p_1 p_2 - q_1 q_2, \\
\dot{q}_2 &= -2p_2 + \tfrac{1}{2} p_1^2 - \tfrac{1}{2} q_1^2.
\end{aligned} \quad (6.5)$$

Obviously the origin $(0, 0; 0, 0)$ is an equilibrium point. Is it L-stable? To answer the question, observe that for any fixed constant τ the functions

$$p_1 = \sqrt{2}\,\frac{\sin(t - \tau)}{t - \tau}, \qquad p_2 = \frac{\sin 2(t - \tau)}{t - \tau}$$

$$q_1 = -\sqrt{2}\,\frac{\cos(t - \tau)}{t - \tau}, \qquad q_2 = \frac{\cos 2(t - \tau)}{t - \tau} \qquad (6.6)$$

satisfy the Eqs. (6.5) for all $t \neq \tau$. If $\tau \neq 0$, the initial values of these solutions can be obtained by letting $t = 0$:

$$p_1' = \sqrt{2}\ \frac{\sin \tau}{\tau}\ , \qquad p_2' = \frac{\sin 2\tau}{\tau}\ ;$$

$$q_1' = \sqrt{2}\ \frac{\cos \tau}{\tau}\ , \qquad q_2' = -\ \frac{\cos 2\tau}{\tau}\ .$$

The distance of this point from the origin is $\sqrt{3\tau^{-1}}$. Therefore, by choosing τ as a sufficiently large positive number, we can find a solution (6.6) which at time $t = 0$ starts as close to the origin as we please. What happens to the solution as t increases? At any time t, $0 < t < \tau$, its distance from the origin is $\sqrt{3}\ (\tau - t)^{-1}$, which becomes infinite as $t \to \tau$. We conclude that the origin is not L-stable.

EXERCISE 6.1. Verify that the Eqs. (6.6) furnish a solution of the system (6.5)

EXERCISE 6.2. For each of the following Hamiltonians, where $m = 1$, the point $p = 0$, $q = 0$ is an equilibrium point. Determine in each case whether the point is L-stable. (a) $\frac{1}{2}(p^2 + q^2)$; (b) $\frac{1}{2}(p^2 - q^2)$; (c) $\frac{1}{2}p^2 - \cos q$. [In the last case, show that $\frac{1}{2}p^2 + (1 - \cos q)$ remains constant, so that p^2 and $(1 - \cos q)$ must remain small if they are so initially.]

*EXERCISE 6.3. Show that $H(p_1, \ldots, p_m; q_1, \ldots, q_m)$ remains constant in time if $p_1, \ldots, p_m$; $q_1, \ldots, q_m$ is a solution of the system (6.1). Suggestion: show that $dH/dt = 0$.

EXERCISE 6.4. Use the conclusion of the preceding exercise to show that the origin is always a point of

L-stability for a Hamiltonian of the form $\sum c_k p_k^2 + \sum d_k q_k^2$, $c_k > 0$, $d_k > 0$, $k = 1, \ldots, m$.

EXERCISE 6.5.   Show that if $c_k > 0$, $k = 1, \ldots, m$ but $d_k < 0$ for some value of k, then the origin cannot be L-stable for the Hamiltonian of the preceding exercise.

7.   INFINITESIMAL STABILITY

In 1941, Wintner,* writing about the concept of L-stability, said:

> This definition of stability seems to be the natural one. Actually, it is not natural at all. In fact everything that is known from Poincaré's geometrical theory of real differential equations and from the parallel, though more difficult, theory of surface transformations points in the direction that condition (ii)** cannot be satisfied except in highly exceptional cases. Even in the restricted problem of three bodies, not a single solution is known to be stable.

Writing thirty-five years later, it is easy to be wise. Within the last few years, as a result of the work of Kolmogoroff and his school, it is now established that some of the libration points are indeed L-stable. But their methods are far beyond the scope of this book, and we turn aside to look at an easier question.

Classically it was, and remains, customary to substitute for L-stability another concept of stability which is much

---

* *The Analytical Foundations of Celestial Mechanics*, Princeton University Press, 1941, p.98.

** This is the ϵ-δ condition described in Sec. 6.

easier to handle. The basic idea is this. Consider once again the system (6.1) and an equilibrium point $p_1^0, \ldots, p_m^0; q_1^0, \ldots, q_m^0$. Let $p_k = p_k^0 + \epsilon_k$, $q_k = q_k^0 + \eta_k$ represent a solution of the system (6.1) which is "near" the equilibrium solution. Expand each of the derivatives $\partial H/\partial p_k$, $\partial H/\partial q_k$ through terms of the first order in ϵ_k and η_k around the point $p_1^0, \ldots, p_m^0; q_1^0, \ldots, q_m^0$. Because the first partial derivatives themselves vanish at the point, we obtain

$$\frac{\partial H}{\partial p_k}(p_1, \ldots, p_m; q_1, \ldots, q_m)$$

$$= \sum_{l=1}^{m} \left(\frac{\partial^2 H}{\partial p_k \partial p_l} \right)_0 \epsilon_l + \left(\frac{\partial^2 H}{\partial p_k \partial q_l} \right)_0 \eta_l + \text{terms of higher order;}$$

$$\frac{\partial H}{\partial q_k}(p_1, \ldots, p_m; q_1, \ldots, q_m)$$

$$= \sum_{l=1}^{m} \left(\frac{\partial^2 H}{\partial q_k \partial p_l} \right)_0 \epsilon_l + \left(\frac{\partial^2 H}{\partial q_k \partial q_l} \right)_0 \eta_l + \text{terms of higher order;}$$

the subscript 0 indicates that the second derivatives are to be evaluated at the point. Now let

$$a_{kl} = \left(\frac{\partial^2 H}{\partial p_k \partial p_l} \right)_0, \qquad b_{kl} = \left(\frac{\partial^2 H}{\partial p_k \partial q_l} \right)_0,$$

$$c_{kl} = \left(\frac{\partial^2 H}{\partial q_k \partial p_l} \right)_0, \qquad d_{kl} = \left(\frac{\partial^2 H}{\partial q_k \partial q_l} \right)_0. \tag{7.1}$$

Then the Eqs. (6.1) become (we write the equations for $\dot{p}_k$ first)

$$\dot{\epsilon}_k = -\sum_{l=1}^{m} c_{kl}\epsilon_l - \sum_{l=1}^{m} d_{kl}\eta_l$$

$$\dot{\eta}_k = \sum_{l=1}^{m} a_{kl}\epsilon_l + \sum_{l=1}^{m} b_{kl}\eta_l, \tag{7.2}$$

provided the terms of higher order can safely be dropped. By this we mean, somewhat optimistically, that a solution of the exact Eqs. (6.1) which starts sufficiently near the equilibrium point will mimic the behavior of that solution of the *linear* system (7.2) which starts in the same position relative to $(0, \ldots, 0; 0, \ldots, 0)$. With this in mind, we define the point $(p_1^0, \ldots, p_m^0; q_1^0, \ldots, q_m^0)$ to be *infinitesimally* stable for the system (6.1), if the origin is L-stable for the linear system (7.2). The infinitesimal stability of equilibrium solutions is what is studied in classical mechanics under the theory of "small" oscillations.

Is the optimism justified? Is a point which is infinitesimally stable also L-stable? We examine the system (6.5) for which the origin is *not* L-stable. Let $p_k = \epsilon_k$, $q_k = \eta_k$ and drop the terms which are not linear. The resulting system is simply

$$\dot{\epsilon}_1 = -\eta_1, \qquad \dot{\epsilon}_2 = 2\eta_2, \qquad \dot{\eta}_1 = \epsilon_1, \qquad \dot{\eta}_2 = -2\epsilon_2. \tag{7.3}$$

The solution of this is

$$\epsilon_1 = A \cos t - C \sin t,$$

$$\epsilon_2 = B \cos 2t + D \sin 2t,$$

$$\eta_1 = C \cos t + A \sin t,$$

$$\eta_2 = D \cos 2t - B \sin 2t,$$

where $(A, B; C, D)$ are the initial values of $(\epsilon_1, \epsilon_2; \eta_1, \eta_2)$. It is easy to check that $(\epsilon_1^2 + \epsilon_2^2 + \eta_1^2 + \eta_2^2)^{1/2} = (A^2 + B^2 + C^2 + D^2)^{1/2}$. It follows that if the solution starts within ϵ of the origin, it remains within ϵ of the origin. This is more than enough to guarantee L-stability for the linear system.

We have shown that a point which is stable according to the classical theory need not be stable according to the desirable criterion of L-stability. Nevertheless, the classical method has its uses and we shall discuss it at length.

EXERCISE 7.1.   Show that for the examples described in Exs. 6.2–6.5, the two definitions of stability give consistent results.

8. THE CHARACTERISTIC ROOTS

We have seen that the problem of infinitesimal stability of an equilibrium point leads to the study of linear systems (7.2). The traditional method of solving such systems is to look first for solutions of the form $\epsilon_k = A_k e^{\lambda t}$, $\eta_k = B_k e^{\lambda t}$. Substitution into Eqs. (7.2) leads to the linear system*

$$\sum (-c_{kl} - \lambda)A_l + \sum (-d_{kl})B_l = 0$$
$$\sum a_{kl}A_l + \sum (b_{kl} - \lambda)B_l = 0. \tag{8.1}$$

Denote by A, B, C, D, respectively, the $m \times m$ matrices (a_{kl}), (b_{kl}), (c_{kl}), (d_{kl}) defined by (7.1) and let I denote the

------

* In this section $\sum$ means $\displaystyle\sum_{l=1}^{m}$.

$m \times m$ identity matrix. The matrix of coefficients is then

$$\mathfrak{M} = \begin{pmatrix} -C - \lambda I & -D \\ A & B - \lambda I \end{pmatrix}.$$

If the determinant of the coefficients is not zero, then the system (8.1) has only the solution $A_k = 0$, $B_k = 0$, $k = 1, \ldots, n$. A non-trivial solution can be guaranteed if the determinant vanishes. This means that λ must satisfy the equation $|\mathfrak{M}| = 0$. If we multiply each of the first m rows by (-1), we get

$$\begin{vmatrix} C + \lambda I & D \\ A & B - \lambda I \end{vmatrix} = 0. \tag{8.2}$$

The left-hand side is a polynomial in λ of degree $2m$. Its roots are called the *characteristic roots* of the system (7.2). We shall prove that for systems (7.2) whose coefficients originate in a Hamiltonian, as indicated by (7.1), the polynomial is even. This means that *if λ is a characteristic root, so is $-\lambda$.*

Observe first that, according to (7.1), the matrices B and C are transposes of one another, so that (8.2) may be written

$$\begin{vmatrix} C + \lambda I & D \\ A & C^T - \lambda I \end{vmatrix} = 0. \tag{8.3}$$

Since, by (7.1), $A = A^T$, $D = D^T$, we may transpose the determinant on the left to obtain

$$\begin{vmatrix} C^T + \lambda I & A \\ D & C - \lambda I \end{vmatrix} = 0.$$

Now interchange the last m rows with the first to get

$$\begin{vmatrix} D & C - \lambda I \\ C^T + \lambda I & A \end{vmatrix} = 0.$$

Finally, interchange the last m columns with the first. We obtain

$$\begin{vmatrix} C - \lambda I & D \\ A & C^T + \lambda I \end{vmatrix} = 0.$$

This shows that if λ satisfies (8.3), so does $-\lambda$, and the proof is complete.

EXERCISE 8.1.   Find the characteristic roots of the system (7.3).

EXERCISE 8.2.   Show that 0 is a characteristic root if and only if the Hessian of the Hamiltonian vanishes at the equilibrium point, that is, if and only if

$$\begin{vmatrix} A & B \\ C & D \end{vmatrix} = 0.$$

9.   CONDITIONS FOR STABILITY

Suppose we are testing an equilibrium point $(p_1^0, \ldots, p_m^0; q_1^0, \ldots, q_m^0)$ for stability. We start by looking at infinitesimal stability. To avoid complications which arise in the general case, but not in the problems we consider, let it be supposed from now on that *the $2m$ characteristic roots $\lambda_1, \ldots, \lambda_{2m}$ are distinct*. This means, in particular, that none of them can be zero since the associated polynomial is even; if one λ were zero, two of them would be.

It is now easy to prove: the origin is L-stable for the system (7.2) or, what is equivalent, *the point $(p_1^0, \ldots, p_m^0;$*

$q_1^0, \ldots, q_m^0)$ *is infinitesimally stable if and only if all the numbers* λ_k *are pure imaginary.*

First suppose that some λ_k has a non-zero real part. Then one of the numbers λ_k, $-\lambda_k$ has a positive real part. The general solution of (7.2) contains terms of the form $e^{\lambda_k l}$, $e^{-\lambda_k l}$. One of these becomes infinite in magnitude as $t \to \infty$. Therefore the origin cannot be infinitesimally stable.

Conversely, if all the λ_k are pure imaginary, let $\lambda_k = i\mu_k$, μ_k be real. The general solution of (7.2) is of the form

$$\epsilon_k = \sum_{l=1}^{2m} A_{kl} e^{i\mu_l l}, \qquad \eta_k = \sum_{l=1}^{2m} B_{kl} e^{i\mu_l t}.$$

Therefore $|\epsilon_k| \leqslant \sum_{l=1}^{2m} |A_{kl}|$, $|\eta_k| \leqslant \sum_{l=1}^{2m} |B_{kl}|$ for all time. If we choose the sums on the right to be small, then the solution remains small for all time. This completes the proof of the theorem stated in the second paragraph.

How does this help with the problem of L-stability? Only to this extent. It was shown* by Liapounov that if the Hamiltonian has continuous partial derivatives of the third order, then a point cannot be L-stable unless it is infinitesimally stable. The condition on the Hamiltonian will be met in our problems. Therefore we can conclude that *if any* λ_k *has a non-zero real part, the point is not L-stable.* On the other hand, if the system is infinitesimally stable, that is, all the λ_k are pure imaginary, no conclusion about L-stability can be drawn without further investigation. This is demonstrated by the examples given in Sec. 7.

---

* See, for example, L. Cesari, *Asymptotic Behavior and Stability Problem in Ordinary Differential Equations*, New York: Academic Press, Inc., 1963, p.93.

In the next section we shall investigate the stability of the five libration points. To avoid distracting digressions, we ask the reader to verify some computations in the following exercises. The notation is that of Secs. 12 and 13 in Chap. 2. We let

$$s = (1 - \mu)\rho_1^{-3} + \mu\rho_2^{-3},$$

$$A = \frac{\partial^2 \Phi}{\partial \xi^2}, \qquad B = \frac{\partial^2 \Phi}{\partial \xi \partial \eta}, \qquad C = \frac{\partial^2 \Phi}{\partial \eta^2}.$$

*EXERCISE 9.1. Show that

$$A = 1 + 2s - 3\eta^2\left[(1 - \mu)\rho_1^{-5} + \mu\rho_2^{-5}\right],$$

$$B = 3\eta\left[(1 - \mu)(\xi + \eta)\rho_1^{-5} + \mu(\xi - 1 + \mu)\rho_2^{-5}\right],$$

$$C = 1 - s + 3\eta^2\left[(1 - \mu)\rho_1^{-5} + \mu\rho_2^{-5}\right].$$

*EXERCISE 9.2. Keeping in mind that $\eta = 0$ at the libration points L_1, L_2, L_3 and that $\rho_1 = \rho_2 = 1$ at L_4 and L_5, verify the following table of the values of A, B, C at the libration points:

	A	B	C
L_1	$1 + 2s$	0	$1 - s$
L_2	$1 + 2s$	0	$1 - s$
L_3	$1 + 2s$	0	$1 - s$
L_4	$\frac{3}{4}$	$\frac{3}{4}\sqrt{3}(1 - 2\mu)$	$\frac{9}{4}$
L_5	$\frac{3}{4}$	$-\frac{3}{4}\sqrt{3}(1 - 2\mu)$	$\frac{9}{4}$

*EXERCISE 9.3. With each of the libration points, we shall associate two numbers x_1, x_2 which are the

roots of the quadratic $x^2 + (4 - A - C)x + (AC - B^2)$. Prove that at L_4 and L_5 the numbers x_1 and x_2 are both negative if and only if $27\mu(1 - \mu) < 1$.

*EXERCISE 9.4.   For the libration points L_1, L_2, L_3, the quadratic of the preceding problem becomes $x^2 + (2 - s)x + (1 + 2s)(1 - s)$. Prove that if $s > 1$, not both roots can be negative at the same time.

*EXERCISE 9.5.   At each libration point, $\partial\Phi/\partial\xi$ must vanish. Use this to show that at each of L_1, L_2, L_3

$$(1 - \mu)(\rho_1 - \rho_1^{-2})\frac{\xi + \mu}{\rho_1} + \mu(\rho_2 - \rho_2^{-3})\frac{\xi + \mu - 1}{\rho_2} = 0.$$

Show that at L_1 this can be written

$$\rho_1(s - 1) = \mu(1 - \rho_2^{-3}),$$

and at L_2, L_3

$$\rho_1(1 - s) = \mu(1 - \rho_2^{-3}).$$

Since $\rho_2 > 1$ at L_1 and $\rho_2 < 1$ at L_2 and L_3, conclude that $s > 1$.

*EXERCISE 9.6.   Combine the preceding exercises to conclude that $x^2 + (4 - A - C)x + (AC - B^2)$ cannot have two negative roots at L_1, L_2, L_3. At L_4 and L_5 it has two negative roots if and only if $27\mu(1 - \mu) < 1$.

10.   THE STABILITY OF THE LIBRATION POINTS

Recall from Sec. 7 that the equilibrium points of the restricted three-body problem are five in number and have coordinates $(-q_2^0, q_1^0; q_1^0, q_2^0)$ where (q_1^0, q_2^0) are the coordinates of the corresponding libration points in the ξ-η coordinate system. In order to test these points for stability, we must compute the coefficients defined by (7.1) and the determinant which occurs in (8.3). If we start with the Hamiltonian in the form

$$\tfrac{1}{2}\left(p_1^2 + p_2^2\right) - (q_1 p_2 - q_2 p_1) + \tfrac{1}{2}\left(q_1^2 + q_2^2\right) - \Phi(q_1, q_2),$$

it is easily verified that (8.3) becomes

$$\begin{vmatrix} \lambda & -1 & 1 - \Phi_{11} & -\Phi_{12} \\ 1 & \lambda & -\Phi_{12} & 1 - \Phi_{22} \\ 1 & 0 & -\lambda & 1 \\ 0 & 1 & -1 & -\lambda \end{vmatrix} = 0, \qquad (10.1)$$

where the subscripts indicate partial differentiation with respect to the variables q_1 or q_2. Since $q_1 = \xi$, $q_2 = \eta$, evaluation of the determinant yields

$$x^2 + x(4 - A - C) + AC - B^2 = 0, \qquad (10.2)$$

where $x = \lambda^2$ and A, B, C have the same meaning as in Exs. 9.1–9.6.

Now if the points are to be L-stable, it is necessary that all the λ be pure imaginary. Therefore both roots x of (10.2) must be negative. According to Ex. 9.6, this is never possible for L_1, L_2, L_3. Hence the points are unstable. The same exercise shows that L_4, L_5 are unstable when $27\mu(1 - \mu) \geqslant 1$.

Therefore the only possible cases of stability left are L_4 and L_5 when $27\mu(1 - \mu) < 1$, that is, $\mu < .03852$. In the major examples of interest to astronomers, this condition on μ is satisfied and many years of observation indicate that the points are L-stable. A theoretical proof of stability has appeared only recently, thus settling an important question of long standing. The proof is due to the Russian mathematician Leontovich who used advanced methods devised by Kolmogoroff and Arnold.*

EXERCISE 10.1 Verify the derivation of Eqs. (10.1) and (10.2) from the given Hamiltonian. Confirm that the four roots λ are distinct and pure imaginary when $27\mu(1 - \mu) < 1$.

---

* See *Russian Mathematical Surveys*, XVIII (1963), p. 13, Example 4.

INDEX

1976 MAA Basic Library List

Series -- Do not break